CONSISTENCY OF FOODSTUFFS

CONSISTENCY OF FOODSTUFFS

by

TOSHIMARO SONE

D. REIDEL PUBLISHING COMPANY

DORDRECHT-HOLLAND

SHOKUHIN NO NENCHOSEI

First edition published by Korin Book Publishing Co., July 1966; second edition July 1969.
Translated from the Japanese by Sachio Matsumoto

Library of Congress Catalog Card Number 78–183368

ISBN 90 277 0219 5

Printed in The Netherlands

PREFACE

It has become increasingly clear that more attention is being paid to the physical chemistry and texture of foodstuffs as developments in food technology proceed more rapidly. Rheology, which has been skillfully introduced into polymer technology during recent years, should also be of direct relevance to food technology.

The consistency of foodstuffs not only influences the processing of materials e.g. transport, mixing, etc., but it is also important for controlling the quality of food products. An older friend of mine, Dr T. Takano, and I have published a book *Flow and Transport in Food Technology*, in which we explained the concepts of rheology in the first part, and discussed the technological applications in the latter part.

As far as text books on the subject of rheology of foodstuffs are concerned, we have only a few at the present, for example, *Foodstuffs, their Plasticity, Fluidity and Consistency* edited by Dr Scott Blair, and *Agricultural Rheology* written by Dr Scott Blair and Dr Reiner. Although they are now classical works, there is clearly a difference between the social conditions of European countries and Japan.

Since rheology is a science for describing the relationship between stress, deformation and time, foods should offer much of interest to the field of rheology because of their characteristic behaviour. It seems, however, that food acceptability depends more on taste than on consistency, that is, in general people do not seriously consider the consistency of foods on their palate. On the other hand, a number of test machines have been designed and used for many years for measuring the characteristic properties of foods in order to establish reliable standards of quality. However, we have to appreciate that such an approach may have little in common with rheology.

In this book, the term consistency is used to make clear the sense of the term texture which is widely used for the comprehensive assessment of foodstuffs. It is not possible to systematize the characterization of foodstuffs, though it may be necessary to explain systematically the consistency of foodstuffs from the standpoint of Korin's series of books on Foods.* This situation can be compared with ancient times when people could not understand the correlations which exist between the planets in the sky. It is hoped that this book may offer the reader a number of topics in the field of food rheology.

Foods are of course a motive force for providing vital energies to all human life, and we have a wide range of materials and products in foodstuffs. This book deals mainly with the consistency of the primary processed products, so that those who are

* The translator notes that Korin Publishing Company is trying to complete a series of text books on food science and technology.

specialists in the field of food technology may perhaps feel that something is missing in the various chapters. I hope, however, that if a comparative study is made of all chapters the reader will find several subjects of interest.

The author would be very happy, if this book would promote and stimulate further discussion on the various subjects mentioned therein.

TOSHIMARO SONE

PREFACE TO THE ENGLISH EDITION

It is my great honour and pleasure to have my small book introduced to a wide area of the world through this edition.

Food, clothing and habitation form the requisites of our life. Especially, food is essential to life. In every country in the world the traditional and distinctive types of foods have been maintained and provided over many centuries. During recent years, processing technology has been developed for manufacturing a wide range of foods, and meanwhile many investigations on the texture and consistency of foodstuffs have been made by physicists, chemists, biologists, psychologists and technologists. However, the results of investigations in this, and allied, fields are scattered over a wide spectrum of journals. Although one experiences much inconvenience as result of this tendency, it is inevitable that there will be overlaps between many other fields and food science and technology. It also seems that solution of many food problems in the future will depend upon the success of research activities in various basic fields. In this respect, I would be very happy, if this small book will stimulate readers to find an interest in food rheology and allied problems.

I am very much indebted to Dr P. Sherman, Unilever Research Laboratory, Welwyn, Hertfordshire, England, for reading the translated manuscript and making many valuable suggestions. Also, he has encouraged me to publish this edition. Finally, I would thank my friend Dr S. Matsumoto, University of Osaka Prefecture, Japan, who translated the whole text into English.

TOSHIMARO SONE

Meguro, Tokyo
May 1971

TABLE OF CONTENTS

CHAPTER 1

WHAT IS FOOD CONSISTENCY?

1.1. Texture and Consistency

Foodstuffs which consist of carbohydrates, fats, proteins, and a small amount of vitamins and minerals, are the materials for supplying us with vital energies for growth, metabolism, vital activities, and so on, through digestion and assimilation in the body.

We eat many foodstuffs everyday as part of our staple diet (e.g. rice, wheat, bread, potato, etc.), subsidiary diet (e.g. animal proteins, vegetables, etc.), luxury foods (e.g. sweets, cakes, drinks, fruits, seasonings, etc.). Foodstuffs are generally cooked because one requires a satisfactory taste to be transmitted through the sense organs. In other words, foodstuffs are required to provide some sensory elements for satisfying the human palate. The various elements in the sensory problem are generally colour, taste and flavour. In order to complete the sensory aspect of foods, however, it is necessary to add an element of form to the other properties. The form referred to means the microscopic structure of foods including their appearance and shape, so that it is better to use the term texture in this context rather than the term form.

The technical term texture, which means interweave in Latin, has been used for many years to describe a close or rough texture in pictures or in paintings in addition to its usage for foodstuffs. Webster's dictionary defines the term texture as the dimensional and morphological properties of the microscopical components of a body or material. The consistency of liquids and fluids may also be represented by the term texture. It is clear that texture is one of the technical terms which may be used for describing the physical qualities of foodstuffs.

The physical qualities which have been expressed objectively and subjectively, should be represented by pure quantities, or by physical meanings. Thus, first, the sensory response to stimulation must be defined using physics and psychology. Second, it is necessary to apply objective measurements in order to clarify the physical qualities. A systematized approach for study of the first case is the so-called psycho-rheology or psychophysics, and the second case belongs to rheology.

In the field of food science, however, one approach cannot be separated from the other, because the texture of foodstuffs is more or less assessed by sensory means. For example, we have a number of sensory expressions such as 'toughness', 'smoothness', 'body', 'crispness', and so on. When attempting to replace these expressions by physical quantities, it may be not possible to represent the sensory responses by simple measurements of viscosity, or elasticity, or energy of breaking. Sensory expressions generally define judgements made in the mouth or by the finger, so that an analysis must be made to obtain a correlation between the stimulus and the sensory response, and then the results are substituted for the physical qualities. On the other hand, the rheological approach is based on a study of the relationship between stress, strain, and time for the materials which are being examined.

It is doubtful, whether the sensory evaluation is made by applying a small stress or with a constant time-scale, although small stresses or small deformations, and a Newtonian time-scale, are generally required to investigate the rheological behaviour of the sample. For example, honey or a very concentrated aqueous sugar solution does not deform under the influence of a local pressure, but jelly can easily be cut by a spoon. However, when one compares the stiffness of honey with that of jelly, one may judge that honey is much softer.

According to the science of mechanics, honey is a viscous fluid, and jelly is an elastic body. Viscous fluids deform spontaneously under gravity, or the rate of deformation is proportional to the value of the stress. Such a property can be expressed by the term viscosity, which is one of the material constants for fluids. However, jelly easily retains its shape under gravity for a long period of time, and the deformation of jelly under stress is perfectly elastic in an ideal case, i.e. the deformation occurs instantaneously under the influence of a stress, and it is proportional to the value of the stress. Such a property can be described by the term elasticity, which is also one of the material constants for elastic bodies.

Jelly is easily cut with a spoon, because the stress introduced partially by the spoon is so much larger than the yield value of the elastic deformation that the structure of jelly is broken down. A comparison of the stiffnesses of honey and jelly may be likened to a comparison of viscosity with elasticity. It is just as if one compared size with the time scale, i.e. it is a comparison of different dimensions. The physical definition of the texture of foodstuffs with a single dimension is quite difficult, since it is not possible to describe the texture of foodstuffs by a simple physical quantity. This problem will be discussed later.

The technical term gel-strength is useful for describing the mechanical behaviour of a jelly. Gel-strength is actually determined by means of the pressure required for a plunger to penetrate the surface of a jelly, so that both elastic deformation and breakdown of the sample under the shearing stress are contained in the term gel-strength. Accordingly, such a term may strongly resemble the sensory response on the tongue or by the teeth. Kramer [1] has pointed out that materials whose mechanical behaviour can be described by the term gel-strength should be regarded as the subject matter of texture studies, but that fluids, e.g. sauces, which flow spontaneously under gravity, are not part of texture studies. When sauce is flowing out from a vessel, one can identify the flow properties, e.g. viscous, thick, or thin, visually. Therefore, Kramer has explained, measurements of the firmness of such solid foods as meats, beans, etc. are relevant to a study of their texture, because their firmness has to be judged by the sense organs in the mouth or by the fingers.

In view of the above, Kramer has attempted to find a correlation between the standards of sensory assessment and the appropriate rheological terms, as shown in Table 1.1. This table suggests that the term Newtonian or non-Newtonian describes the nature of those materials which deform under a stress which is less than 1.0 G, and that the term texture should be used for describing the mechanical properties of those materials which deform under a stress which is greater than 1.0 G. Thus, it

TABLE 1.1

Relation of 'texture' to other sensory and rheological terms

Psychological or sensory terms	Rheological or physical terms (gravitational force)		
	Up to 1.0 gravity		Greater than 1.0 gravity
	Newtonian	Non-Newtonian	
Sight ... flow or spread	Viscosity	Consistency	Texture
Feel ... mouth or finger			
Taste and smell		Flavor	

seems that the term texture is deeply concerned with the sensory description of the firmness of solid-like foodstuffs. The rheological term 'yield value' or 'elasticity' is not defined in Table 1.1, although the term consistency denotes both rheological and psychological expressions of the flow properties of non-Newtonian fluids.

Although the term consistency is employed in Table 1.1, it seems that this term represents the apparent viscosity of non-Newtonian fluids, such as concentrated suspensions or emulsions. In general, the term consistency is used to describe the viscous properties of liquid materials, but Reiner [2] has shown that even solid-like materials behave like viscous fluids after long loading times, e.g. it can be seen in a creep curve of concrete. Thus, for foodstuffs the term consistency is defined in terms of both the micro- and macroscopic quantities of mechanics such as viscosity, elasticity, or plasticity, irrespective of whether the food is fluid or a solid.

1.2. Consistency and Taste of Foodstuffs

If the hardness of metal is compared with that of glass, one may judge that glass is somewhat harder. With metallic materials, one may also judge that iron appears harder than aluminium. When the hardness of glass is compared with that of polystyrol resin, it may seem that polystyrol resin is somehow softer than glass. As in the above examples so also in foodstuffs one has an impression of firmness as well as colour, taste, or flavour. Accordingly, a foodstuff with an abnormal firmness would not be acceptable.

The taste of foods is derived partially from those components of foods which dissolve in saliva during mastication in the mouth, but the sensory evaluation of the toughness of foods while chewing is also a contributory factor. For example, a good taste is conferred on meat by a suitable combination of meat components such as fibrin, meat juice, etc., and of mechanical properties such as toughness, rigidity, etc. One can spread butter on bread but not shortening, although both are fat products. The spreadability properties of low melting point-fats are different from those of butter. One may also judge that the oiliness of butter in the mouth is somehow different from that of the oil component of butter. The unique consistency of butter is closely associated with its taste. Foodstuffs are generally dispersed systems in which the components are blended together. The taste components exist in various forms in food, e.g. in solution in the oil phase or in the aqueous phase, in a crystalline state, in colloidal dispersion, etc. Each component participates in the formation of the whole structure of food. Therefore, it follows that different food consistencies will give rise to different sensory assessments of taste in the mouth, since different rates will be involved in bringing the taste elements from the structural components of food into contact with the sense organs in the mouth.

From the above considerations, it is possible to suggest that some of the descriptive sensory expressions we have used for foodstuffs such as 'toughness', 'touch', 'brittleness', 'body', etc., are closely related to the sensory terms used to describe the tatste of foodstuffs. The taste of 'Udon' which is a form of noodles used in Japan, depends very much on its toughness, i.e. the cooked Udon gradually loses its toughness as it cools in the dish*, so that it is preferable to use a protein-rich wheat flour to prepare the dough of Udon. 'Yohkan'** is made by mixing sugar together with 'An'† for a long time in order to induce a higher consistency. 'Kamaboko'†† is generally assessed on the basis of its toughness or body. These examples emphasize that the taste of foodstuffs is greatly influenced by a synergistic effect with the consistency.

From the historical point of view, ice cream was prepared originally by mixing broken pieces of ice with milk, but now it is manufactured as a frozen product. Ice crystals are dispersed together with air-cells in present-day ice cream. This dispersion state provides a unique consistency and palatability to the ice cream, and if a number of rough ice crystals exist in the system, then the ice cream has a strange taste. We have many other examples in addition to the above, such as boiled rice, mash potatoes, jelly, etc.

Although it is easy to identify a close relationship between the taste and the consistency of foods, as shown by the above examples, a quantitative analysis has not yet been fully established.

Recently, Kishimoto [3] has reviewed his texture studies on Kamaboko, and he suggests that the sensing element for the palate is formed by an interrelation between the gel-strength and the toughness, or between the toughness and the shortness.

* The translator notes that the Japanese people express such a phenomenon as the so-called 'stretch'; it may actually be due to the Udon gradually swelling in the soup.
** Sweet jelly of beans.
† Bean jam (paste).
†† Fish jelly.

1.3. Classification of Consistency

When the term consistency is used for assessing the quality of foodstuffs, its meaning changes according to the nature of the raw materials, or to the structure and the form of the products. The changeable meaning of the term consistency is also widely applied to describe the influence of both raw materials and products on sensory, assessment.

An article [4] tried to classify the term consistency of foodstuffs, as follows: (1) the sensation in the mouth, (2) foods with a simple form, e.g. syrup, candy, etc., (3) foods in the emulsion or suspension state, (4) foamed cakes, e.g. marsh-mallow, etc., (5) gelled foods, such as pectin gel, (6) bread, or cakes made from wheat-flour, (7) chocolate, (8) scientific principles for measuring the texture, (9) measurements of the firmness of meat, (10) measurements of the texture of fish muscle, (11) effects of bean tissue upon the chemical composition, (12) tissue of the cooked potatoes, (13) factors affecting the texture of plant tissue, and (14) strength of eggs. In his article, however, any correlation between the above independent items is not fully determined.

A book [5] on the rheological aspects of foodstuffs has been edited by Scott Blair and written with the cooperation of some European rheologists. In it eight items are independently discussed. They are (1) starch, (2) cereals, (3) milk, ice cream mix and similar products, (4) consistency of butter, (5) rheology of cheese, (6) rheology of honey, (7) rheology of complex food products, and (8) psychorheology of food-stuffs.

Matz [6] has proposed the following classifications for the texture of foodstuffs.

(1) Classification based on the raw materials of foodstuffs; e.g. meat, fish, fruit, etc.

(2) Classification based on the tissue-components of foodstuffs; e.g. muscle, fatty component, etc.

(3) Classification based on the chemical compositions of foodstuffs; e.g. protein, starch, etc.

(4) Classification based on the physical structure of foodstuffs; e.g. fibroid, gel state, emulsion, etc.

(5) Classification based on sensory assessment: e.g. oily, hard, etc.

However, the above classifications are presented as wholly independent of each other, and in some cases an attribute or property is quite different according to the food of its origin. For example, in the case of classification (1), juiciness of fish and of fruit belong to different categories. Furthermore, in the case of classification (3), most foodstuffs are complex mixtures of various components of protein, starch, and so on, and it is rather difficult to pinpoint the main component in such complicated systems.

On the basis of these criticisms, Matz has presented yet another classification:

(1) liquid (viscosity is a main property),

(2) gel (elasticity is a main property),

(3) fibroid foodstuff (fibroid tissue and its strength are important),

(4) aggregated structure of large cells (fruit, potato, etc.),

(5) fatty foodstuffs,

(6) brittle foodstuffs,

(7) glassy foodstuffs,

(8) spongy foodstuffs,

(9) complex foodstuffs (various combinations of the above properties).

It seems to the author that the granular or powdery state should be included in the above classification. Another important problem here is that some characteristics of foodstuffs may be altered by the manufacturing process, so that it is always necessary to give attention to any property changes during processing.

References

[1] Kramer, A.: 1964, *Food Technol.* **18**, 304.

[2] Reiner, M.: 1961, *Deformation, Strain and Flow*, H. K. Lewis & Co., London, p. 158.

[3] Kishimoto, A.: 1965, *Zairyo* (J. Japan Soc. for Testing Materials) **14**, 264.

[4] Anonyme: 1960, *Texture in Food, Soc. Chem. Ind. Monograph* **7**.

[5] Scott Blair, G. W.: 1953, *Foodstuffs: Their Plasticity, Fluidity and Consistency*, Interscience Publishers, New York.

[6] Matz, S. A.: 1962, *Food Texture*, The Avi Publishing Co. Inc.

CHAPTER 2

OBJECTIVE MEASUREMENTS OF CONSISTENCY

2.1. Objective Expressions of Consistency

In order to define the consistency of materials in physical terms, the mechanics of mass such as viscosity, elasticity, viscoelasticity, etc., are combined with non-linear phenomena such as breaking, yield value, etc. Viscosity and elasticity are the prototype principles of rheology, and rheological properties of materials can be explained using a combination of these principles. The author will describe viscosity first in order to interpret systematically the term consistency.

In physics, viscosity is defined by the ratio of shearing stress to rate of shear, but the measurement must be made at a constant temperature. Shearing stress is the force which acts in opposite directions on both sides of an ideological face in the body. In the case of liquids, the rate of shear which is brought about by the internal friction between the molecules is proportional to the value of the shearing stress. Figure 2.1a, shows a plot of rate of shear $(\dot{\gamma})$ against shearing stress (S) and the gradient of the linear graph denotes the viscosity or viscosity coefficient in those cases where the rate of shear is not influenced by time. The above type of liquid is called a purely viscous liquid or a Newtonian liquid.

It is quite rare to have foodstuffs which behave as Newtonian liquids and it is not possible to attach a constant coefficient to the relationship between shearing stress and rate of shear in cases such as those shown in Figures 2.1b, c, and d.

If S dyne cm^{-2} is the shearing stress and $\dot{\gamma}$ s^{-1} rate of shear, then the viscosity η for a Newtonian liquid is then

$$\eta = S/\dot{\gamma} \quad [\text{poise}] \tag{1}$$

In the case of the non-linear curves b, c, and d in Figure 2.1, the respective relations can be given by

$$\eta^* = (S - \tau)/\dot{\gamma} \tag{2}$$
$$\eta^* = (S)^n/\dot{\gamma} \tag{3}$$
$$\eta^* = (S - \tau)^n/\dot{\gamma} \tag{4}$$

where η^* is the apparent viscosity, but it is not defined in poise. The fluid characterized by Equation (3) is called a de-Waele, or pseudoplastic body, and the fluid characterized by Equation (4) is called a plastic body. When $n=1$ for the plastic body, the apparent viscosity can be described by Equation (2), where τ is the yield value, and this type of the fluid is called a Bingham body. It is actually difficult to find a

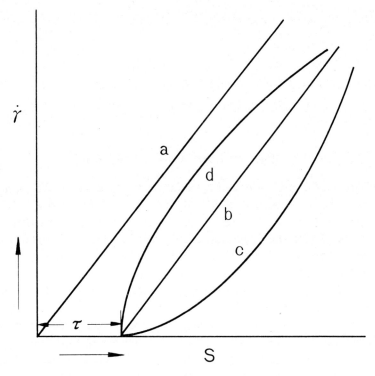

Fig. 2.1.   Viscous liquid.

pure Bingham body. The yield value signifies that shear flow occurs when the shearing stress exceeds the value of τ, but that shear flow cannot be achieved when the shearing stress is less than τ.

The value of τ indicates the critical value of stress for a perfectly elastic deformation, so that a permanent deformation never occurs below this critical value. To obtain an accurate estimate of the critical value is, however, rather difficult, because the flow curve plot of S against $\dot{\gamma}$ actually shows three yield values for a non-Bingham body, as shown in Figure 2.2.

On a molecular level the Bingham plastic body or the plastic body can be regarded as possessing a network structure and the yield value corresponds to the stress required to destroy the network structure. The network may not break instantaneous-ly, but instead the structure breaks down gradually according to some probability factor, so that, in general, the yield value cannot be recognized as a distinct scale as can the pure Bingham body.

Metzner [1] has subdivided non-Newtonian fluids into two groups viz. time-independent fluids and time-dependent fluids, and he has also divided them further according to their respective flow curves, as follows:

Time-independent fluid:

(1) Bingham plasticity or plastic fluid,

(2) pseudoplastic fluid,
(3) dilatant fluid.
Time-dependent fluid:
(1) rheopectic fluid,
(2) thixotropic fluid.
Dilatant fluids in the above classification behave contrary to pseudoplastic fluids, i.e. in a dilatant fluid the increase of shear rate decreases with increasing shearing stress, whereas for pseudoplastic fluids the rate of shear increases more than proportionately with increasing shearing stress. The behaviour of a dilatant fluid can be expressed by

$$S = \eta^* (\dot{\gamma})^n \tag{5}$$

where the term n must be taken as below zero or above unity.

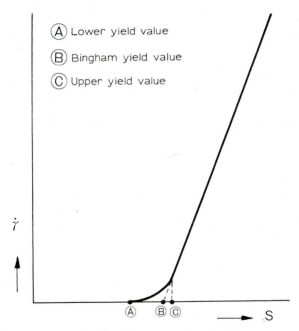

Fig. 2.2.   Three types of yield value.

Time-dependent fluids are characterized by the fact that the rate of shear changes with time under a constant shearing stress. For example, a starch granules-in-water suspension requires a rather large force for agitating the system, but the required force gradually decreases as the agitation continues. A material which shows the above flow behaviour is called a rheopectic fluid. The term thixotropy is used to describe the flow properties when the viscosity is easily reduced by agitation, and then it gradually recovers on standing.

Let us suppose that the molecules of a viscous fluid can be represented by a model which consists of a number of parallel thin plates, as shown in Figure 2.3. We can consider the friction between the plates as viscosity. It is understandable from this model that the heat due to the internal friction between the plates is lost. Accordingly, viscous flow is a thermo-dynamically irreversible process, and the viscosity produces heat due to the work resulting from the application of an external force. Therefore, entropy is formed increasingly throughout the irreversible process.

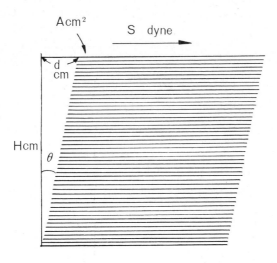

Fig. 2.3.   Shear deformation.

An energy loss caused by internal friction cannot be seen with an elastic body subjected to a small deformation. For a perfectly elastic body the deformation under constant stress is wholly recovered by removal of the stress. This response should occur instantaneously on the time scale. In general, however, most materials show a continuation of the deformation with time following an instantaneous strain under constant stress. Such behaviour can be expressed by a mechanical model consisting of a combination of a spring element (for a perfect elastic body) and a dashpot element (for a perfect viscous fluid). One of the basic mechanical models, the so-called Maxwell element, is serially composed of a spring and a dashpot, and the other, the so-called Kelvin or Voigt element, is composed of both the above elements in parallel. If one end of the Maxwell element is pulled against the other end by a constant stress S, and then if the strain resulting from the stress is kept at constant, the spring element should instantaneously expand with a constant strength, and the dashpot should then stretch slowly. Therefore, the strength of the expanded spring gradually decreases with increasing time, while the energy is scattered and lost through the dashpot element, so that the stress is also gradually relaxed. Such a relaxation of the initial stress S after a period of time t can be expressed by the equation

$$\frac{dS}{dt} = E\frac{d\gamma}{dt} - \frac{E}{\eta}S,\tag{6}$$

where E is the elasticity, η is the viscosity, and γ is the strain. If we write that $\eta/E = \tau_{rel}$, one may derive that τ_{rel} corresponds to a period of time for reducing the initial stress S to $1/e$. This time scale is called the relaxation time. When τ_{rel} is much larger than t, i.e. immediately after the deformation resulting from the application of stress, the Maxwell model behaves as an elastic body. On the other hand, when t is much larger than τ_{rel}, a Maxwell system behaves like a viscous fluid. In a Kelvin or Voigt system the spring element gradually expands in parallel with the dashpot element under the constant stress, so that the stress does not relax, but instead the strain is induced continuously as a retardation phenomenon. The correlation between stress, strain, and time can be defined by the equation

$$S = E\left(\gamma + \frac{\eta}{E}\frac{d\gamma}{dt}\right).\tag{7}$$

We can also write that $\eta/E = \tau_{ret}$, which denotes the retardation time. For a Kelvin or Voigt system τ_{ret} means that when the stress is removed from the system, the deformation recovers gradually by $1/e$ in a period of time τ_{ret}. Therefore, this system also reveals either elastic or viscous properties, due to the ratio of τ_{ret} to t. The systems described above are called viscoelastic bodies.

The concepts of relaxation time and retardation time are shown only by simple models such as Maxwell and Kelvin or Voigt elements, so that these models can be applied to materials whose stress or strain on a logarithmic scale can be related linearly to the time scale. These models should be regarded as first-order approximations. In general, the viscoelasticity of most materials is represented by a combination of so many elements that their relaxation times or retardation times will be distributed around a mean value on the time scale. Stress or strain appears in the material in response to a stimulus, and the response to the stimulus from the molecular structure of the material becomes more complex as the structure becomes more complicated. Thus, it follows that it may be difficult to obtain interrelationships between the structure of a food and its spectrum of relaxation times or of retardation times. Especially in the case of materials whose mechanical properties are characterised by a time dependence, the simple mechanical models described above are only useful for obtaining elementary data about the structure of the material.

Foodstuffs generally consist of so many components such as soft fats, and high polymers like fibrous materials, protein, and starch which are masticated easily, and decomposed to low molecular weight materials by enzymatic actions, that it is very difficult to consider the mechanical properties of foodstuffs purely on the basis of information about the consistency of each component. The homogeneity of foodstuffs is an important subject for study with respect to their consistency. It may be quoted as a special feature of foodstuffs that temperature, moisture, and unit operations such as kneading, mixing, milling, etc., also play a big part in determining the mechanical properties of the final product.

2.2. Measurement of Consistency

2.2.1. VISCOSITY

The apparatus for measuring the viscosity of liquid-foods can be classified roughly into three types such as (1) capillary, (2) falling sphere, and (3) rotational cylinder, respectively.

(i) *Capillary Type*

Viscosity measurement using the very popular capillary viscometers is based on Poiseuille's law, as follows;

$$\eta = (\pi P r^4 / 8vl)\, t \qquad\qquad (8)$$

where η is the viscosity in poise, P is the pressure to the liquid sample in g cm^{-2}, r is the radius of the capillary in cm, v is the flux of the sample in cm^3 s^{-1}, l is the length of the capillary in cm, and t is the measured time in seconds.

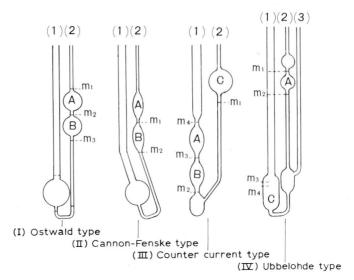

Fig. 2.4.   Types of capillary tube viscometer.

Some typical capillary viscometers are the Ostwald type, Cannon-Fenske type, counter current type, and Ubbelohde type, as shown in Figure 2.4, in which the graduation marks m_1, m_2, m_3, and m_4 are used for introducing a fixed volume of the liquid sample into the viscometers, and for measuring the flux time of the sample. A fixed volume of sample is introduced to ensure a constant shearing stress in the viscometer, and the graduation marks for measuring the flux time are necessary to ensure a constant rate of shear under the influence of the constant average shearing stress.

The principle of the method most commonly employed is that if measurements of both the flux time t_2 and the density ϱ_2 of the sample are made, the viscosity η_2 can be calculated by comparison with a standard liquid whose density is ϱ_1, flux time is t_1, and viscosity is η_1, as follows;

$$\eta_1/\eta_2 = \varrho_1 t_1/\varrho_2 t_2 \qquad (9)$$

For example, one can easily determine the viscosity of a sample by measuring its flux time in the viscometer, because accurate information is available about the density and viscosity of water at various temperatures, and the flux time of water can easily be determined in the viscometer.

One of the pipette type viscometers is used by the Japanese Institute for Technology to establish the export standard of sodium alginate. This equipment operates at a shearing stress of 27.5 dyne cm^{-2}. It is possible to obtain flow curves over a wide range of shearing stresses or rates of shear from the measurements made during a single run using the Maron-Krieger viscometer [2], as shown in Figure 2.5, which

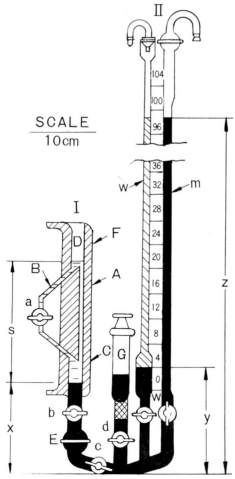

Fig. 2.5.   Maron-Krieger viscometer.

is operated by producing a pressure. The apparatus consists of a capillary section and a manometer section in which mercury or a less dense liquid than mercury creates the pressure which induces flow of the sample through the capillary tube. If the height of the mercury meniscus is denoted as z, which decreases continuously with time, $z - z_0 = h$ and this corresponds to the change of shearing stress in the capillary tube, where z_0 is the position of hydrostatic balance in the manometer tube.

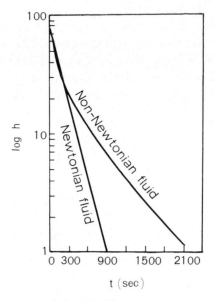

Fig. 2.6.   Flow curve.

Figure 2.6 shows the correlation between $\log h$ and time t with both Newtonian and non-Newtonian fluids. In the case of Newtonian fluids, the slope in Figure 2.6 is related to the reciprocal viscosity of the fluid by

$$\frac{d \log h}{dt} = -\frac{\beta}{\eta} \tag{10}$$

where β is the apparatus constant. The above Equation (10) is, however, not linear for the case of a non-Newtonian fluid, so it is understandable that the viscosity of the fluid depends on the value of the shearing rate.

Because the viscosity of a non-Newtonian fluid depends upon shearing stress or rate of shear, it can only be expressed as an apparent value. An equation which has been derived independently by Saal and Koens, and Maron and Krieger, shows that the change of maximal shearing stress at the wall of the capillary tube S_m induces the change of shear rate D, i.e. the relationship between the above condition $D = f(S_m)$ and the apparent fluidity of a non-Newtonian fluid $\phi_a (= 1/\eta_a)$ is given by the equation

$$\frac{f\left(S_m\right)}{S} = \phi_a + \frac{1}{4} \cdot \frac{d\phi_a}{d \ln S_m}.$$ (11)

If we write the plastic viscosity of the fluid as η_{pl}, and the yield value by f, the following relation can be derived,

$$Q = \frac{R^4 S}{SL\eta_{pl}} \cdot \left\{1 - \frac{4}{3} \cdot \frac{2fL}{RS} + \frac{1}{3}\left(\frac{2fL}{RS}\right)^4\right\}$$ (12)

where Q is the flux of the fluid, R is the radius of the capillary tube, S is the shearing stress, and L is the length of the capillary tube. The above relation has been called the Buckingham-Reiner equation [3, 4].

The viscosity of milk during the induction period of coagulation by added rennet has been investigated by G. W. Scott-Blair *et al.* [5] using a capillary type viscometer, as shown in Figure 2.7, in which A is the glass tube whose length is about 40 cm, and the inner diameter is 0.7 cm. The upper end of the glass tube is connected to a funnel, and it is also fixed to the apparatus by means of a cork stopper. The lower end of the glass tube is connected by a rubber tube to a capillary tube (7 cm length, 0.04 cm diameter) whose end is placed in a beaker which has its base 0.1–0.2 cm below the end of the capillary tube. Another glass tube also leads from the beaker to the outside of the apparatus in order to keep the pressure inside the beaker at atmospheric pressure.

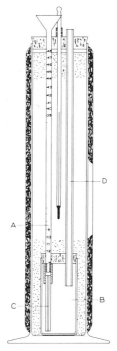

Fig. 2.7.   Thin tube viscometer.

The glass tube A, which has graduations extending to the upper end from 10 cm above the lower end, is immersed in a cylindrical glass vessel together with another glass tube, and one can easily read the graduations of both the tube A and a thermometer. The cylindrical vessel is filled with water whose temperature is the same as that of the cheese vat. Fifteen ml of milk sample is then introduced into the funnel from the cheese vat about 5 min before it begins to coagulate. Thus, the sample will reach the beaker 1–2 min after the introduction of the sample through the glass tube A and the capillary tube, and then one can measure the change of height h of the sample meniscus in the tube A using the graduations and stopwatch, or a metronome during 10–15 s intervals, and then one may plot $\log h$ against time t.

The relationship between $\log h$ and t is linear for a Newtonian fluid, but for milk plus rennet the graph has a tendency to show curvature. This viscometer belongs to the Maron-Krieger category.

(ii) *Falling Sphere Viscometer*

Design of the falling sphere viscometer is based on Stokes' law. When a sphere falls through a liquid, the viscosity of the liquid η can be calculated by the following equation, which is considered to cancel the wall effect of the vessel

$$\eta = \frac{d^2 (\varrho_0 - \varrho)\, gt}{18l} \cdot \left\{ 1 - 2.104\, \frac{d}{D} + 2.09 \left(\frac{d}{D}\right)^3 \right\} \tag{13}$$

where d and D are the diameters of the sphere and the cylindrical vessel, ϱ_0 and ϱ are the densities of the sphere and the liquid, respectively, l is the distance through which the sphere falls in a period of time t, and g is the gravitational constant. If a preliminary experiment is made using a standard liquid whose viscosity is precisely known, Equation (13) can be modified to a simple form, as follows;

$$\eta = K\, (\varrho_0 - \varrho)\, t \tag{14}$$

where K is the shape constant of the apparatus, and t is the time taken by the sphere to fall through a distance l in the standard liquid whose viscosity and density are η and ϱ, respectively. Therefore, one may easily determine the viscosity η_s of a sample using the relation

$$\frac{\eta}{\eta_s} = \frac{(\varrho_0 - \varrho)\, t}{(\varrho_0 - \varrho_s)\, t_s} \tag{15}$$

where ϱ_s is the density of the sample, and t_s is the time taken by the sphere to fall through the sample. Hoeppler's falling sphere viscometer may be quoted as a typical one in which the cylindrical vessel is inclined to the horizontal, so that the sphere rolls along the vessel wall as it passes through the liquid sample. The ascending sphere viscometer operates on the same principle as the falling sphere type, e.g. a bubble is frequently utilized for measuring the viscosity of transparent liquids with reasonable accuracy.

The falling sphere viscometer can only be used with those liquids which satisfy Stokes' law, and which adhere to the surface of the sphere. Therefore, this type of viscometer cannot be used with non-Newtonian fluids.

(iii) *Rotational Viscometer*

When a rotating cylinder is immersed in a liquid, either the rotational speed of the cylinder at constant torque or the torque of the rotation at constant speed is influenced by the viscosity of the liquid. This is the operating principle of the rotational viscometer. It is widely used for quality control of materials in manufacturing processes because viscosity can be measured more easily with the rotational viscometer than with the capillary type.

When using a coaxial cylinder viscometer the viscosity of a Newtonian liquid is given by the equation

$$\eta = \frac{T}{4\pi L \Omega} \left(\frac{1}{R_1^2} - \frac{1}{R_2^2} \right) \tag{16}$$

where T is the torque of the bob, Ω is the angular velocity of the rotation, L is the effective length of the bob, and R_1 and R_2 are the radii of the bob and the cup. When the radius of the cup is infinity, $1/R_2^2$ will be zero in Equation (16). If the sample possesses a yield value f, the following relation can be derived for the flow curve and yield value;

$$\eta^* = \frac{T}{4\pi L \Omega} \left(\frac{1}{R_1^2} - \frac{1}{R_2^2} \right) - \frac{f}{\Omega} \ln \frac{R_2}{R_1} \tag{17}$$

The above equation is generally known as the Reiner-Rivlin formula [7].

We have many types of rotational viscometers such as Stormer's type, Couette's type, McMicheal's type, or Green's type. The clearance between the bob and the cup in the McMicheal type viscometer is so narrow that this viscometer is only suitable for measuring viscosity at the higher rates of shear. On the other hand, the Brookfield type viscometer, which is generally utilized as an apparatus for quality control, has a wide clearance between the bob and the cup, and it is possible to change the bob for others which have a different size or shape. Other instruments similar to the Brookfield's viscometer are the B-type viscometer (Figure 2.8) manufactured by the Tokyo Keiki Co. Ltd., and the viscotester which is manufactured by Rion Co. Ltd. Brookfield's viscometer will measure viscosities up to 8×10^4 poise, although the measurable range of viscosity depends on the actual model employed.

Cone-and-plate viscometers are a kind of rotational viscometer. As shown in Figure 2.9, the cone and the plate are precisely located within a co-centric axis, where either the cone or the plate can be rotated co-centrically, while the other is fixed. The cone angle θ against the plate is as small as $0.5-4°$. Cone-and-plate viscometers give a uniform rate of shear throughout the sample, so that the dependence of viscosity upon rate of shear appears uniform even with non-Newtonian flow. The relationship between the angular velocity of rotation Ω and the torque T

Fig. 2.8.   Brookfield's type viscometer.

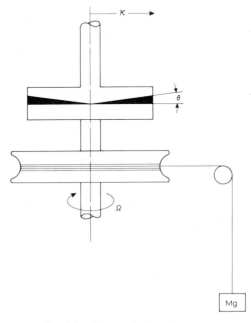

Fig. 2.9.   Cone-and-plate viscometer.

for the cone-and-plate viscometer can be given by

$$T = \frac{2\pi\eta\Omega R^3}{3\theta} \tag{18}$$

where R is the diameter of the sample on the plate, and θ is the cone angle against the plate, so that the viscosity of the sample may easily be calculated by the above equation.

In view of the practical implications for measuring the consistency of foodstuffs, mention should also be made of some typical instruments such as the Brabender amylograph-viscograph, Cone industrial viscometer with recorder, V.I. Viscograph, etc. They are designed according to the same principles as the Brabender Viscocorder which consists of a mixing screw and a sample cup, as shown in Figure 2.10, for measuring the torque of mixing under a constant velocity of mixing.

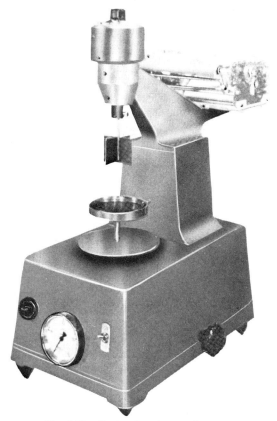

Fig. 2.10. Brabender visco-corder.

2.2.2. ELASTICITY

(i) *Coaxial Type Elastometer*

The structure of elastic liquids such as a dilute solution of agar or gelatine is of much interest, and coaxial cylinder rheometers have been used frequently to investigate this problem. For many years Schwedoff's apparatus [8] (Figure 2.11) has been employed as a coaxial cylinder rheometer in which the rigidity of the sample can be determined by measuring the torsion moment of the bob. The bob is suspended by means of a torsion wire whose torsion constant is k, and when the sample is introduced into the clearance between the bob and the cup it can be sheared by torsion of the bob through an angle θ which is brought about by twisting the upper end of the torsion wire through an angle ϕ. Therefore, the torsion moment of the bob M is given by

$$M = k\,(\phi - \theta) \quad (\text{dyn cm}^{-1}) \tag{19}$$

and the rigidity G of the sample is calculated by the equation

$$G = \frac{k}{4\pi L}\left(\frac{1}{r_1^2} + \frac{1}{r_2^2}\right)\cdot\frac{(\phi - \theta)}{\theta} \quad (\text{dyn cm}^{-2}) \tag{20}$$

where L is the effective length of the bob. A perfect elastic body deforms without any

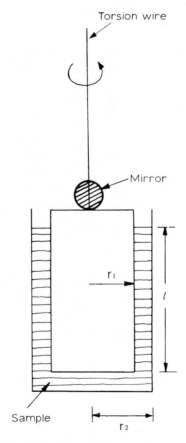

Fig. 2.11.   Schwedoff's apparatus.

loss of the internal friction, so that the deformation is wholly recovered when the stress is removed. Such a phenomenon is similar to the compression or expansion of an ideal gas.

(ii) *Compressive Type Elastometer*

Saunders and Ward's apparatus [9] can be used to measure the elastic compression and recovery of a not too concentrated gel. Figure 2.12 shows a schematic diagram of this apparatus. The sample is introduced into the thick glass tube whose upper end is connected with the air reservoir through the thin glass tube, and the lower end of the thick glass tube is continued to a capillary tube manometer into which mercury is filled in order to measure the strain of the sample. The air reservoir applies air pressure to the sample by means of a compresser, the air pressure being read by use of the manometer. The rigidity of the sample G can be obtained by the relation

$$G = (PR^4)/(8Lr^2h) \tag{21}$$

where R and r are the radii of the thick glass tube and the capillary tube manometer, L is the length of the sample in the thick glass tube, P is the pressure, and h is the height of mercury in the capillary manometer tube.

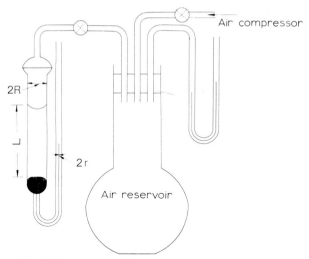

Fig. 2.12.   Apparatus for measuring the elasticity of gels.

Ueno *et. al.* [10] have made measurements on the creep of alkali treated gelatine using a Saunders and Ward's type apparatus, and they found that the dependence of elasticity of the gelatine sample upon concentration can be divided into three regions such as a square law region, a dispersion region, and a glassy state region, respectively. Scott Blair *et. al.* [11] have modified Saunders and Ward's apparatus with particular respect to the U-tube sample vessel and the gradient manometer tube. The modified apparatus has been used to measure the gel-elasticity of milk during the initial stages of coagulation.

(iii) *Measurements of Tensile Strength*

Very little has been written about the tensile strength of foodstuffs although knowl-
edge of how the structure influences the mechanical strength of foodstuffs is useful
for obtaining information about factors affecting both manufacturing processes
and consistency. When stress is applied to a body, the induced force can be sub-
divided into two types of stress, i.e. one as perpendicular, and the other as shearing
stress between planes which are considered to be located parallel to the direction of the
force. It appears that shearing stress is related to the spreadability of paste or
chocolate when they are used for covering or coating purposes, but the shearing
stress is always smaller than the tensile stress. Charm [12] has investigated tensile
strength with an apparatus, as shown in Figure 2.13, in which the sample is slowly
pressed out from the perpendicular pipe under a force which is too small to cause
flow in the sample. When the weight of the sample extruded from the pipe exceeds
the tensile strength, part of the sample breaks off from the sample column. Therefore, if one measures the weight of the broken off sample and the cross-sectional area
of the sample column one can calculate the tensile strength of the sample S_t using
the following relation

$$S_t = W/A, \tag{22}$$

where W is the weight of the fractured sample, and A is the cross-sectional area of the
sample column at which the fracture occurs. It is also possible to calculate the critical
diameter R of the sample column at fracture using shearing stress C, the weight of
the sample in the pipe G, the density of the sample ϱ, and the length of the pipe L,
as follows;

$$R = (2C)/\varrho. \tag{23}$$

The tensile strength of gel-like viscous foods, such as mayonnaise, ketchup, etc.,
is about twice as large as the shearing stress, and the sample is fractured by a force
which is larger than the shearing stress. This shearing stress develops in two directions,
each at forty-five degrees to the main plane of stress, as shown in Figure 2.13b. In the
case of foodstuffs it is not possible to identify with the naked eye the slip plane
(Lueder's line) which occurs in metals, although this plane actually exists in foodstuffs also.

When a viscous foodstuff is extruded from the pipe, the waist line can be seen in
the sample column just before fracture because of the dead weight. However, fracture
does not always occur after the waist line has appeared; if one stops extruding the
sample, it sometimes remains without any fracture. This phenomenon emphasizes
that the sample exhibits strain hardening, because the sample is slowly extruded from
the pipe without flow. In practice, butter and margarine, which show yield values
in their flow curves, also exhibit the above phenomenon.

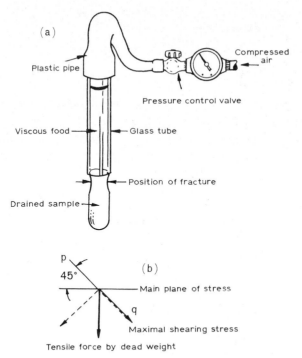

Fig. 2.13.   Apparatus for measuring the tensile strength.

2.2.3. MEASUREMENTS OF VISCOELASTICITY

(i) Static Methods

When one wants to measure the viscoelasticity of foodstuffs such as dough, noodles, fish jelly, ice cream, etc., it is advisable to use various techniques such as stretch under dead weight, stretch and/or recovery of cylinder-like samples on a mercury bed, etc. The tensile machine [13] (Figure 2.14) is also applicable. In the tensile machine, the sample is clamped between the chucks C_1 and C_2, and then tensile stress is induced into the chuck C_1 through the sample block by means of a balance mechanism by pulling down the chuck C_2, as illustrated in Figure 2.15. Therefore, the applied tensile stress to the sample can be obtained by

$$W = W_0 \frac{L_2 L_4}{L_1 L_3} \tan \theta. \tag{24}$$

The value θ can be converted into one of sample expansion, so that it is possible to make creep curve measurements on the sample using the tensile machine.

Shear viscosity η can easily be converted to the normal viscosity λ by the relation $\lambda = 3\,\eta$. If deformation of a rod shape sample (d cm) can be measured by a simple bending or stretching, both viscosity λ and elasticity E can be expressed by the same equations, as follows;

$$E = \frac{PL}{Ad} \quad (\text{dyn cm}^{-2}) \tag{25}$$

$$\lambda = \frac{PL}{A\dot{d}} \quad (\text{poise}) \tag{26}$$

where P is the stress in dyne, A is the cross sectional area of the sample in cm^2, and L is the length of the sample in cm, respectively. Therefore, if the shape factor of the cylinder or square rod sample is well defined, it is possible to obtain both their viscosity and elasticity. Any instrument for measuring static viscoelasticity is basically designed on the principles described above.

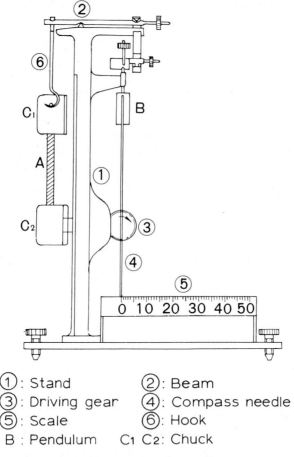

①: Stand   ②: Beam
③: Driving gear   ④: Compass needle
⑤: Scale   ⑥: Hook
B : Pendulum   C₁ C₂: Chuck

Fig. 2.14.   Tensile testing machine.

(ii) *Measurements of Dynamic Viscoelasticity*

When a simple harmonic motion $\theta = \theta_0 \sin \omega t$ is applied to the cup of a coaxial cylinder rheometer, the oscillation induced in the bob is affected by the viscous

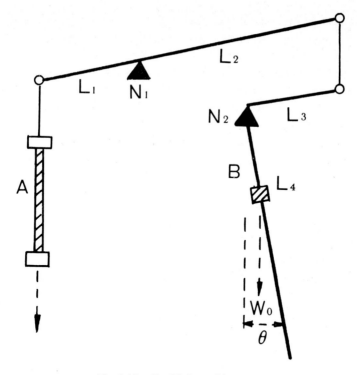

Fig. 2.15.   Equilibrium of leverage.

resistance R and elastic strength K of the sample which is filled into the clearance between the bob and the cup. The phase of oscillation in the bob is shifted negatively, and the amplitude also differs from that of the cup. If the radii of the bob and cup are r_1 and r_2 respectively, and the effective length of the bob is h, one can obtain both the dynamic viscosity η' and the rigidity G of the sample using the following equations;

$$\eta' = \frac{R}{4\pi L} \left(\frac{1}{r_1^2} - \frac{1}{r_2^2} \right) \quad \text{(poise)} \tag{27}$$

$$G = \frac{K}{4\pi L} \left(\frac{1}{r_1^2} - \frac{1}{r_2^2} \right) \quad \text{(dyn cm}^{-2}) \tag{28}$$

It is necessary to determine experimentally both the values of R and K in the above equations. A coaxial cylinder rheometer is usually used for measuring the dynamic viscoelastic properties of liquid materials. In the case of solid foods, we have other possible techniques, e.g. damped free oscillations have been applied to Kamaboko (fish jelly) by Kishimoto et al. [14], and Kimball's apparatus [15] has been used to apply transverse or torsional oscillations to cheese or butter.

2.2.4. INDUSTRIAL METHODS FOR MEASURING CONSISTENCY

Not only food manufacturers but also other industries subject their raw materials and products to various control tests although they have not always carried out basic measurements such as determination of the viscoelastic properties. In practice, each manufacturer has attempted to establish specialized techniques for testing his materials, e.g. test machines for evaluating the qualities of dough or bread. It is now possible to roughly classify these specialized testing machines into groups such as the penetrometer type, the compression type, the consistency meter type, the shear fracture measurement type, etc.

A. *Penetrometers*

Penetrometers have been used widely for a long time as instruments for measuring the hardness of materials because of their cheapness and easy operation. The ASTM has prepared a standard for use of the penetrometer as a testing machine for measuring the consistency of grease. Many kinds of penetrometer are also used to evaluate the consistency of foodstuffs. The following instruments seem to belong to the penetrometer classification in its wider sense, Bloom gelometer for measuring the hardness of various gels, the curd meter and curd tension meter (standardized by the Japanese Society of Technology for Dairy Products) for measuring the gel strength of coagulated milk, micro-penetrometer with a standardized needle as proposed by the AOCS for measuring impact penetration in fats, etc.

The ASTM penetrometer [16] defines the relative hardness of grease using the relationship between the depth of penetration by the needle and the load, so that it is necessary to provide a precise standard for the shape and dimensions of the needle. Figure 2.16 gives these details. The cone part, whose weight is 102.5 ± 0.05 gr, is fixed to the centre axis and the depth of penetration can be measured by means of a dial gauge with an accuracy of $1/10$ mm. The weight of the moving axis is also precisely defined as 47.5 ± 0.05 gr, and the hardness of grease is evaluated by the depth of the penetration into the sample in a period of 5 min at $77° \pm 1°F$. Thelen [17] has proposed an equation for expressing the correlation between the average rate of shear v and the shearing stress S using the depth of the penetration Δp in a period of time t under the total load W, as follows;

$$v = \frac{7.854 \times 10^{-5} \times \Delta p}{t} \tag{29}$$

$$S = \frac{0.098 \times 10^{4} \times W}{0.001571 \times \Delta p - 0.0521}. \tag{30}$$

Haighton [18] has obtained flow curves and yield values for plastic or pseudoplastic edible fats using a simple type of penetrometer, which is shown in Figure 2.17.

Figure 2.18 illustrates the flow curves of edible fats obtained by use of a penetrometer with various cones and different cone angles. From this figure a generalized relationship between the shearing stress S and the rate of shear $\dot{\gamma}$ can be defined as $\log \dot{\gamma} = \log \eta^* + n \log S$, where η^* is the apparent viscosity, and n is the power index.

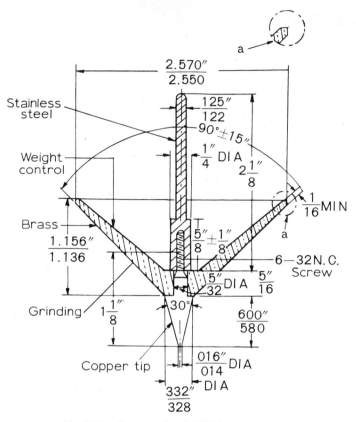

Fig. 2.16. Cone needle of ASTM penetrometer.

It seems to be better to use the yield value rather than apparent viscosity to define the firmness of margarine, butter, and edible fats. If one draws tangent lines from the points B_1 and B_2 on the curve in Figure 2.18, one obtains an intersecting point $\dot{\gamma}_r$ on the ordinate, and also points S_1 and S_2 on the abscissa, in which $S_1 \tan \alpha_1$ is equal to $S_2 \tan \alpha_2$. Therefore, the yield value f can be expressed as

$$f = K \frac{P_m}{\sqrt[n]{S_m}} \tag{31}$$

where K is a constant which depends on the cone angle. In practise, the yield value f is calculated from the relation

$$f = K \frac{W}{p^n} \tag{32}$$

where W is the weight of the cone needle, and p is the depth of penetration. The power index n is generally considered to have a value of 1.6 for margarine, butter, and shortening.

B. *Compression Type Testing Machines*

Delaware's testing apparatus for jelly consists of a syringe and an air compressor for compressing the sample by means of the plunger of the syringe, as shown in Figure 2.19. The air is evacuated to a pressure of 25 pounds/square inch by means of an evacuation valve, and the manometer controls the plunger to obtain a constant rate of compression. The manometer contains carbon tetrachloride, and thirty seconds are required to obtain meniscus differences of 50 cm/carbon tetrachloride. Wittenberger [19] has tried to measure the hardness of pieces of boiled apple, potato, carrot, beet, and vegetables using Delaware's apparatus. With this test apparatus it is possible to attach both physical and mechanical significance to the results.

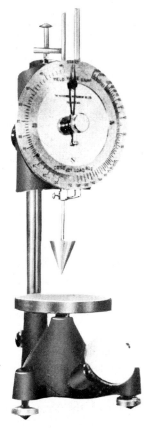

Fig. 2.17.   Cone type penetrometer.

Lovergren [20] has modified Brinel's hardness meter so that it can be used for measuring the hardness of fats, and he has shown that when a ball with diameter D is pushed into the fat sample by a pressure P, the hardness index of the sample H can be given by

$$H = \frac{P(100)}{\frac{\pi D}{2}(D - \sqrt{D^2 - d^2})} = \frac{P(100)}{\frac{\pi}{2}(D^2 - D\sqrt{D^2 - d^2})} \qquad (33)$$

where d is the diameter of the indentation on the sample surface. He has also shown that the reproducibility of the measurement is excellent for the case of $d/D = = 0.20\sim0.25$ using a steel ball with diameter $D = 3/8$ in.

Caffyn and Baron's ball type compressor [21] uses a dial gauge to measure the depth of indentation by a disk-like metal ball. The elastic properties of the sample can also be estimated from recovery measurements in the unloaded condition.

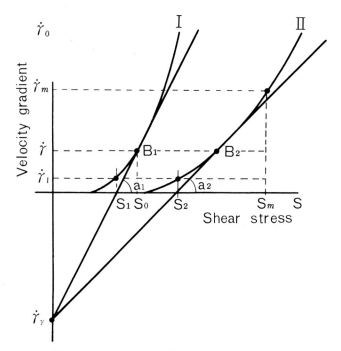

Fig. 2.18.   Flow curve of margarine and edible fat.

TABLE 2.1

Comparison of yield value with the results of sensory assessment for the case of margarine and shortening

Yield value (g cm^{-2})	Assessment
< 50	Very soft, spontaneous flow
50–100	Very soft, but not spreadable
100–200	Soft and spreadable
200–800	Very plastic and spreadable
800–1000	Hard and spreadable
1000–1500	Very hard, a limit of spreadability
> 1500	Much too hard

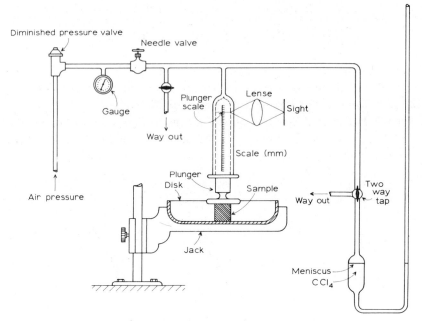

Fig. 2.19.   Delaware's testing apparatus for jelly.

A testing machine suitable for measuring the compression of bread has a square plate (3.6 cm × 3.6 cm), and the softness or sponginess is expressed as the load which is required to compress the bread from 13 mm thickness to 9 mm thickness by means of the plate.

The Technical Centre of the General Foods Co., New York, U.S.A. has designed an instrument for testing the compressibility of gels, as shown in Figure 2.20. The instrument consists of a constant speed motor, a stress transmitor, and a recorder, respectively. The stress transmitor forms a bridge with four strain gauges, and it is connected to a plunger which compresses the surface of the sample with a load of 200 g. Induced strain in the strain gauges corresponds to the gel strength of the sample, and it is recorded automatically by means of a servo-mechanism.

C. *Rotational Type of Testing Machine for Jellies*

In place of the penetrometer, which is mainly used for obtaining yield values of samples, a rotational type testing machine has been used in the United Kingdom for testing jellies, viz. the BFMIRA (British Food Manufacturing Industries Research Association) testing machine [23]. It is an improved version of the BAR (British Association for Research on Cocoa) instrument for jelly. The modified machine is operated as follows; a square plate (2 cm²) is inserted into the jelly sample, and then torque is applied to the axis which is connected perpendicularly to the square plate, in order to produce a torsion of the plate up to 30°. The torque is induced by pouring water into a vessel which is connected through a thread and pulley to the axis, and the strength of the jelly can be represented by the amount of water collected in the vessel. The pouring rate is basically 100 ml/min, although it can be controlled from 20 ml/min to any rate. This testing machine is shown in Figure 2.21.

D. *Shear Press Consistency Meter*

The shear press consistency meter is widely used for testing various important properties of foodstuffs such as the maturity grade of green peas and haricot beans, the hardness of fresh or tinned apples, beet, spaghetti, chicken, and shrimps, the fibriform characteristics of asparagus and meat, etc. One form of this instrument [24] which has been designed by the University of Maryland, U.S.A. consists of six main parts, as shown in Figures 2.22 and 2.23, i.e. (1) a sample box with lattice lid, (2) stainless steel blade which can fit perpendicularly into the lattice lid, (3) proving ring, (4) hydraulic pressure piston, (5) automatically operated hydraulic press, and (6) a recorder. The shear blade is connected with the proving ring so as to remove the error of the pressure transmission.

Fig. 2.20.   Instrument for measuring gel properties.

The sample is placed in the sample box, and then the blade moves down into the sample by means of the piston which operates through the hydraulic pressure transmission. The blade moves the whole distance within a period of time ranging from 15 sec to 100 sec. It is possible to apply pressures up to 5,000 pounds, and the pressure is observed directly by means of the proving ring, or it is recorded automatically.

E. *Testing Machine for Measuring the Softness of Sliced Meats*

Kulwich *et al.* [25] have proposed a method for measuring the softness of cooked meat slices. As shown in Figures 2.24a and b a meat slice is placed on a stainless steel plate which has an hole in the centre, and the sample is covered by an aluminium plate.

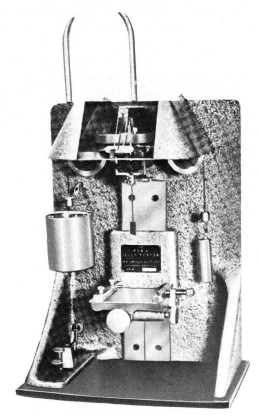

Fig. 2.21.   New FIRA testing machine for jelly.

It is then penetrated by a stainless steel plunger with its lower end in the form of a fine point, while the upper end is a thick cylinder whose diameter is the same as that of the hole in the stainless steel plate. Therefore, in the first stage of the test, the sample is penetrated by a thin needle, and then in the next stage the sample is fractured by punching with a thick cylinder. The diameter of the punched meat is 3/8 in. The plunger can be used in combination with the Instron testing machine, and then the apparatus is able to record automatically the processes of penetration and fracture of the sample. There is a clearance of 0.003 ± 0.001 in. between the plunger and the hole in the stainless steel plate.

The results obtained with chewing gum show that the shear test is reproducible with an error of $2.4\% \sim 2.6\%$, and that the penetration test can be made with fluctuations

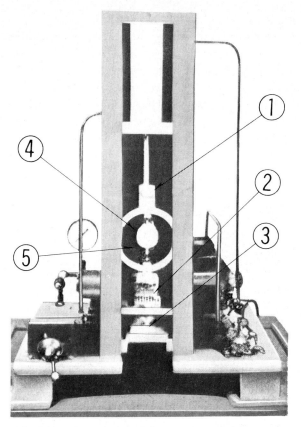

Fig. 2.22.   General view of shear press consistency meter.

of $4.0\% \sim 4.4\%$. In the case of a pork roast sample, one may find a meaningful correlation between the results of this method, the sensory assessment, and the Warner-Bratzler shear test [26], when the shear is applied parallel to the muscle fibres, as shown in Table 2.2.

TABLE 2.2
Penetration test, shearing test, and sensory assessment for roast pork

STE test (lb/0.1 in.)	Correlation coefficient (r)[a]	
	Sensory test	Warner-Bratzler test
Vertical shear for muscular fibrils	−0.61	0.61
Parallel shear for muscular fibrils	−0.72	0.71
Vertical penetration for muscular fibrils	−0.55	0.41
Parallel penetration for muscular fibrils	−0.65	0.51

[a] Judgement was made within 1%.

F. *Texturometer*

We taste the components of food through the process of mastication by the teeth and dilution with saliva. It appears that the stress of mastication plays a big part in the sensory assessment of the firmness of foods, and it also affects the taste stimulus. Procter *et al.*, of MIT, U.S.A., have investigated the mechanism of mastication by combining the mechanical parameters. They employed a model of the human mouth cavity for studying mastication which is based on an apparatus called the Hanau articulator. In this apparatus the model upper jaw moves up and down by means of a driving motor, and the stress fluctuation during the simulated mastication is measured through a strain gauge on the upper jaw with an oscilloscope.

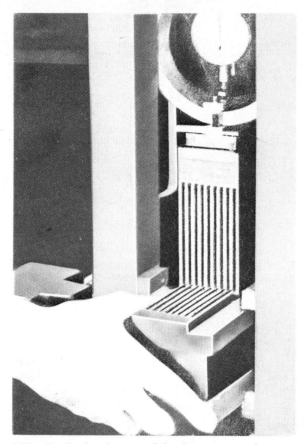

Fig. 2.23.   Detail of testing part of the shear press consistency meter.

This type of testing machine has been improved by Szczesniak *et al.* [27], General Foods Co., New York, U.S.A. Significant points of the modification are as follows; (i) a recorder is used instead of the oscilloscope, (ii) a plunger and plate are employed for the model teeth, (iii) the strain gauge on the upper jaw has been moved to the arm

Fig. 2.24a.   Testing machine for measuring the softness of sliced meat.

which supports the plate, (iv) the rate of mastication can be adjusted, and (v) an accessory unit has been provided for measuring viscosity. Figures 2.25a and b show the modified apparatus and the accessory unit, respectively. Measurement of viscosity using the accessory unit is effected by the cup and the paddle.

Figure 2.26 shows an example of the results obtained by using the texturometer,

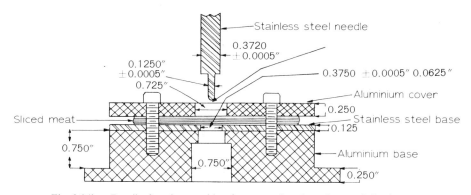

Fig. 2.24b.   Detail of testing machine for measuring the softness of sliced meat.

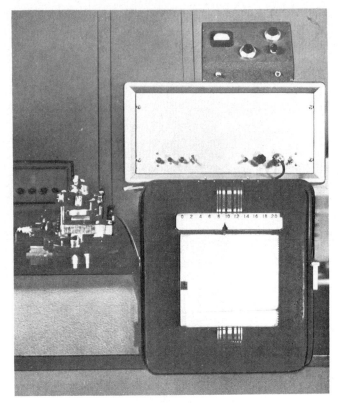

Fig. 2.25a.   Texturometer.

Fig. 2.25b.   Accessory for measuring the viscosity.

in which the ordinate and abscissa correspond to stress and time, respectively. The time scale reads from right to left. In this figure, the peak of the pulse after the first mastication cycle gives the hardness of the sample, and the adhesiveness which is required for pulling out the plunger, is defined by the area A_3. The hardness is represented by the height of the peak per unit input voltage, so that

$$\text{hardness} = \frac{\text{the height of the first peak}}{\text{the input voltages}}$$

Accordingly, it is necessary to prepare samples with a standardized size and shape. Szczesniak suggests that the thickness of the sample should be 1/2 in., and that the minimum distance of separation between the plunger and the plate should be 1/8 in.

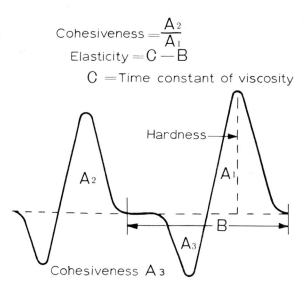

$$\text{Cohesiveness} = \frac{A_2}{A_1}$$
$$\text{Elasticity} = C - B$$
$$C = \text{Time constant of viscosity}$$

Fig. 2.26. An example of the result obtained with the texturometer.

The cohesiveness of the sample is given by the ratio of the area of the second peak to that of the first peak, i.e. A_2/A_1, because this term may be assessed in the mouth as the looseness or tightness of foods. The elasticity E is defined by the relation $E = C - B$, where B is the distance on the abscissa, as shown in Figure 2.26b, and C is the distance on the same axis obtained by using a perfectly non-elastic body as a standard material e.g. clay.

Brittleness can be estimated from the fluctuation pattern of the first mastication peak, and chewiness is calculated as the product of the hardness, cohesiveness, and elasticity. Gumminess is expressed as the product of hardness and the cohesiveness.

Figure 2.27 shows the relative viscosity of diluted cream and maple syrup. The viscosity obtained by the texturometer is only a value relative to the viscosity of the standardized non-elastic clay. An example of brittleness is shown in Figure 2.28a using a dog biscuit, i.e. the fluctuation pattern of the peak represents the brittleness of the sample. White cake shows a high elasticity and softness on the trace from the texturometer, as shown in Figure 2.28b. Texturograms obtained with lemon pudding, which is rather sticky, and with bran flakes, which are relatively hard, are also shown in Figures 2.28c and d.

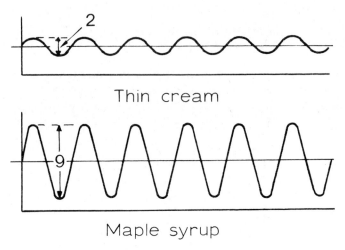

Fig. 2.27.   Patterns of the relative viscosity.

G. *Brabender Farinograph*

This is one of the well-known instruments for testing dough samples (Figure 2.29a), which possesses a couple of screw type agitators in the vessel. The agitators are rotated by a synchronous motor, and the torque induced at the rotor is measured by means of a dynamometer, and it is also recorded on the chart paper. The vessel is surrounded by a thermojacket.

A sample of flour is introduced into the bowl, and it is then mixed with a fixed quantity of water. By this procedure one can compare the consistency of a particular

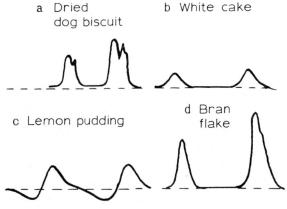

Fig. 2.28.   Results of texturometer tests with various foods.

sample with the consistencies of other samples if the maximal value of the recorder response is adjusted to 500 Brabender Units (BU).

Figure 2.29b shows an example of the results obtained using the Farinograph. From these results one may draw certain conclusions about the properties of dough

(a) The amount of absorbed water, i.e. amount of water required for obtaining 500 BU as the maximal recorder response.

(b) The period of time for attaining 500 BU of the recorder response from the commencement of mixing.

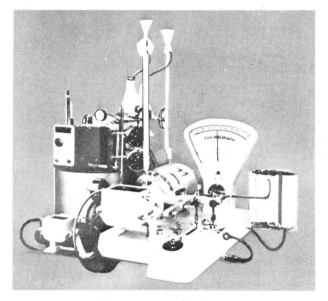

Fig. 2.29a.   Brabender Farinograph.

(c) The period of time for obtaining maximal recorder response from the commencement of mixing.

(d) Durability for mixing, i.e. the period of time that the recorder response keeps position above 500 BU.

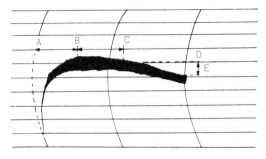

Fig. 2.29b.   Pattern of Farinogram.

(e) Distance of the recorder response from 500 BU at 20 min after the commencement of mixing.

(f) The period of time for decreasing the recorder response from 500 BU to 470 BU.

(g) The BU value at 15 min after attaining the maximal value of the recorder response (500 BU).

(h) MTI (Mechanical Tolerance Index), i.e. the BU value at 5 min after attaining the maximal value of the recorder response; the BU values are measured on the upper edge of the band like response curve.

H. *Brabender Extensograph*

This is a test machine for measuring the extension value and tensile strength of dough, as shown in Figure 2.30a. The dough sample, which is prepared in a specified way, is introduced into the 'cage' of the instrument, and it is then extended to determine the extension length per unit period of time. Figure 2.30b shows a typical response curve. Its height corresponds to mechanical resistance of the sample against extension, and the distance of the curve on the abscissa represents the extension length of the sample. It is generally considered that extension relates to the degree of maturation of the wheat.

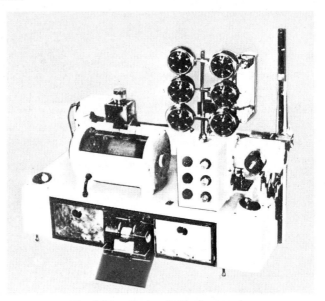

Fig. 2.30a. Brabender Extensograph.

I. *Brabender Amylograph*

The amylograph is an apparatus for recording the consistency change in starch – water or flour – water systems when they are heated. Therefore, the apparatus should also be useful for studying the increase in viscosity of starch during gelation, or for investigating the hydrolysis of starch by α-amylase.

J. *Mixograph*

This instrument possesses a lever and a pin in the mixing bowl, so that one can observe the mechanical resistance to mixing of dough in the bowl. It is possible from the results to estimate the spreadability, stability, and/or water absorbability of dough. In practice, these estimates are utilized to obtain information about the quality of bread.

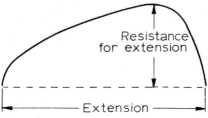

Fig. 2.30b.   Pattern of Extensogram.

2.2.5. MEASUREMENTS OF CONSISTENCY OF POWDERED FOODS

(i) *Definition of the Aggregation State*

A powder constitutes an aggregation state of fine particles which possess various shapes, sizes, and size distributions. The particles exist independently of each other, they make contact with each other, or they combine together, and all these possible conditions are found in the powder system. Therefore, in order to obtain information about the nature of powder systems, it is necessary to have information about the aggregation state of the particles in the powder and also about the nature of the respective particles. The following terms are generally used for defining the static properties of fine particles following aggregation:

a. Apparent Specific Volume – packed volume of powder per unit mass.

b. Apparent Density or Bulkiness – reciprocal number of the apparent specific volume.

c. Void or Porosity – ratio of open volume to the packed volume in the powder system, i.e.

$$\text{void} = 1 - \frac{\text{apparent density}}{\text{density of powder particles}}.$$

d. Void per Unit Weight – open volume of powder per unit mass, i.e.

$$\text{void per unit weight} = \frac{1}{\text{apparent density}} - \frac{1}{\text{density of particles}}.$$

e. Number of Coordinate Particles or Number of Contact Points.

f. Air Content, i.e.

$$\text{air content} = 1 - \frac{\text{density of powder}}{\text{density of powder particles}}.$$

Fig. 2.31.   Brabender Amylograph.

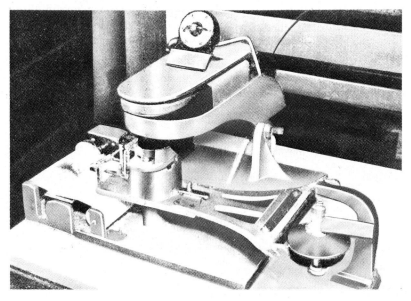

Fig. 2.32.   Brabender Mixograph.

The air content parameter is important for powdered foods. Foods which are prepared by spray drying contain so many bubbles in the particles that the bubbles play a big part in the aggregate structure, and in the physico-chemical properties of the foods.

(ii) Measurement of the Packed State of Particles

A tapping apparatus (Figure 2.33) is widely used for this measurement. The apparatus is designed to measure the volume (apparent density) change in a powder when it is tapped in the powder vessel. During tapping the apparent density of the system will change according to the following equation [28];

$$\varrho_\infty - \varrho_n = A \cdot \exp(-kn) \tag{34}$$

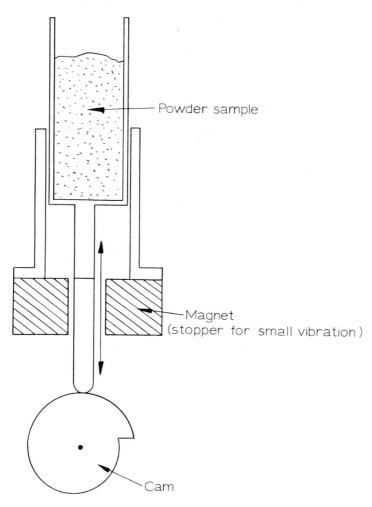

Fig. 2.33.   Tapping apparatus.

where ϱ_n is the apparent density of the system after completing the nth tap, ϱ_∞ is the maximal value of the apparent density, n is the number of taps, and A and k are constants concerned with shape, mass, size, etc. of the particles.

If one considers the voids of the powder using a parameter of volume strain $\gamma=(v_0-v)/v_0$, a relationship between the number of taps n and the volume strain γ can be given by [29]

$$\gamma = \frac{abn}{1 + bn} \quad \text{or} \quad \frac{n}{\gamma} = \frac{1}{ab} + \frac{n}{a}. \tag{35}$$

Suppose that γ equals a, and that the volume of powder becomes v_∞ when $n\to\infty$, than a can be written as $(v_0-v_\infty)/v_0$ which indicates the increase of volume of the powder after an infinite number of taps. The term b in Equation (35) corresponds to $1/N_\tau$, where N_τ is the number of taps required to obtain the half value of the term a.

(iii) *Angle Characteristics of a Powder*

When powder accumulates on a plane, the surface of the accumulated powder tends to slip on the residual layer of the system by gravity, and then this force balances the friction between the powder particles. In this state, the angle of the powder surface to the plane is considered as an angle characteristic of the powder, which appears to be a very important factor affecting the flow properties of the system. We generally quote four angle characteristics of powdered system, as follows.

(a) *Rest Angle*. This is the angle made by the surface of the accumulated powder to the horizontal plane. In practice, as shown in Figure 2.34, it can be obtained by discharging the powder from a box through a hole. The angle formed by the surface of the powder remaining in the bottom of the box is called the introduced angle,

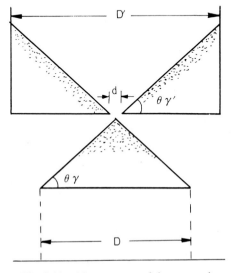

Fig. 2.34.   Measurement of the rest angle.

and the angle of the surface of the powder drained from the box to the plane is called the discharged angle. In the case of dense powders which are prepared so as to have monodispersity without cohesion, the introduced angle is nearly equal to the discharged angle. Powdered foods generally show abnormal flow, as shown in Figure 2.35, and the two angles scarcely agree with each other.

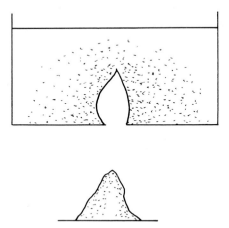

Fig. 2.35.   Abnormality of the rest angle.

(b) *Slide Angle*. One can measure this angle by gradually slanting the plate on which powder is accumulated uniformly and making the powder slide down. The angle of this plate to the horizontal plane is called the slide angle. The cone angle at the bottom of a silo is designed by taking the slide angle into account. An important application of this principle is drainage of powder from a silo under gravity.

(c) *Destruction Angle*. If the side panel of a box which contains powder is removed, part of the powder near the side panel will crumble down under the dead weight of the powder, and finally a kind of precipice is formed on the side surface of the powder. The destruction angle corresponds to the angle of the precipice to the plane. The destruction angle is not always reported in the case of powdered foods.

(d) *Movement Angle*. When powder is introduced into a cylindrical vessel which is rotated about the centre of the cylinder, as shown in Figure 2.36, the surface of the powder will begin to incline, and it will ultimately attain a steady plane of inclination. We denote the steady angle of inclination of the powder in the rotating vessel to the horizontal plane as the movement angle.

(iv) *Measurements of the Coefficient of Internal Friction*

Although there are many methods for measuring the coefficient of internal friction between powder particles, the most convenient one is a direct shearing technique. It seems that the internal friction corresponds to Coulomb's law of friction due to

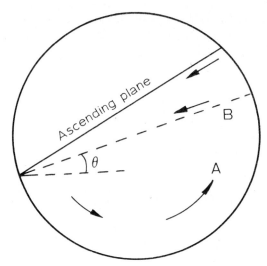

Fig. 2.36.   Measurement of the movement angle.

the relationship between the shearing stress S and the pressure P which acts at right angles to S, i.e.

$$S = \mu P + C. \tag{36}$$

Let us substitute $C = S_0$ at $P = 0$,

$$S = \mu P + S_0. \tag{37}$$

S_0 means the shearing stress when the pressure is zero, so that S_0 corresponds to the force of cohesion between the powder particles.

Figure 2.37 shows the block diagram of an apparatus for measuring the coefficient of internal friction, in which the sample in a cylindrical vessel is pressed by a load, and the induced shearing stress is observed by means of a pressure gauge. The porosity of the sample changes due to a change in the shearing stress, so that the volume of the sample is increased or decreased. This change of volume can be measured by using a dial gauge. The cylindrical vessel containing the sample consists of two parts; an upper part, whose bottom opens to the lower part, is fixed to the apparatus, and the lower part can be slipped out in a side direction from its initial position. So, it is possible to apply a shearing force to the sample at the boundary between both parts of the vessel.

Figures 2.38 and 2.39 show results for shearing tests on dried skimmed milk and sugar, respectively. With increasing shear the volume of skimmed milk decreases and that of sugar increases, which indicates dilatant characteristics [30].

Figure 2.40 shows the correlation between the perpendicular pressure and the shearing force. The slope of the curve corresponds to the coefficient of internal friction, and the intersection of the curve on the ordinate relates to the force of cohesion between the particles.

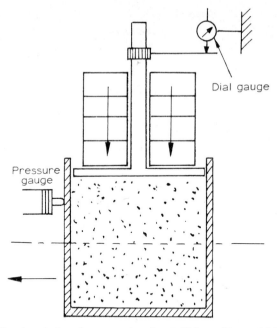

Fig. 2.37. A technique for measuring the coefficient of internal friction.

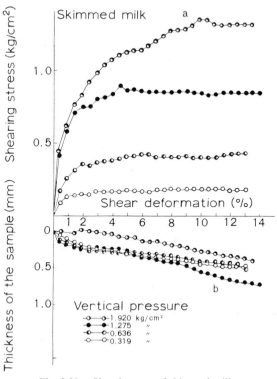

Fig. 2.38. Shearing test of skimmed milk.

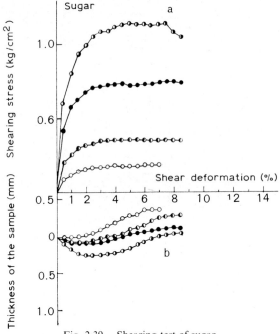

Fig. 2.39.   Shearing test of sugar.

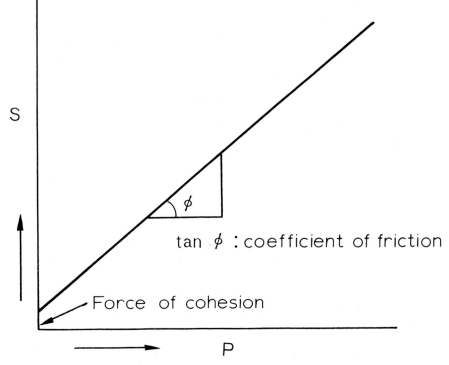

Fig. 2.40.   Relationship between the vertical pressure and the shearing stress in the shearing test.

References

[1] Metzner, A. B.: 1956, 'Non-Newtonian Technology', *Adv. Chem. Eng.* **1**.

[2] Maron, S. H., Kriger, I. M., and Sisko, A. W.: 1954, *J. Appl. Phys.* **25**, 971.

[3] Buckingham, E.: 1921, *Proc. Amer. Soc. Testing Materials* **1154**, 117.

[4] Reiner, M.: 1926, *Kolloid-Z.* **39**, 117.

[5] Scott Blair, G. W. and Oosthuizen, J. C.: 1962, *J. Dairy Res.* **29**, 47.

[6] Höppler, F.: 1933, *Z. Tech. Phys.* **14**, 165.

[7] Reiner, M. and Rivlin, R.: 1927, *Kolloid-Z.* **43**, 122.

[8] Schwedoff, T.: 1889, *J. Phys., Radium* (2) **8**, 341.

[9] Saunders, P. R. and Ward, A. G.: 1954, *Proc. 2nd Intern. Cong. Rheol.* 284.

[10] Ueno, W. and Ohno, I.: 1963, *Zairyo* (*J. Soc. Material Sci. Japan* in Japanese) **12**, 341.

[11] Scott Blair, G. W. and Burnett, J.: 1958, *J. Dairy Res.* **25**, 297.

[12] Charm, S. E.: 1964, *Food Sci.* **29**, 483.

[13] Shimizu, T., Fukawa, H., and Ichiba, A.: 1958, *Cereal Chem.* **35**, 34.

[14] Kishimoto, A. and Hirata, S.: 1963, *Bull. Jap. Soc. Sci. Fish.* **29**, 146.

[15] Kimball, A. L. and Lovell, D. E.: 1927, *Phys. Rev.* **30**, 948.

[16] ASTM Designation, D. 217-52T (1952).

[17] Thelen, E.: 1937, *J. Appl. Phys.* **8**, 135.

[18] Haighton, A. J.: 1959, *J. Amer. Oil Chem. Soc.* **36**, 345.

[19] Wittenberger, R. T.: 1951, *Food Technol.* **5**, 17.

[20] Lovergren, N. V., Guice, W. A., and Feuge, R. O.: 1958, *J. Amer. Oil Chem. Soc.* **35**, 327.

[21] Caffyn, J. E. and Baron, M.: 1947, *Dairyman* **64**, 345.

[22] Szczesniak, A.: 1963, *J. Food Sci.* **28**, 390.

[23] Gaydon, H. A.: 1955, *Food* **24**, 287.

[24] Kramer, A.: 1957, *Food Eng.* **29**, 57; Decker, D. W., Yeatman, J. N., Kramer, A., and Sidwell, A. P.: 1957, *Food Technol.* **11**, 343.

[25] Kulwich, R., Decker, R. W., and Alsmeyer, R. H.: 1963, *Food Technol.* **17**, 201.

[26] Bratzler, L. J.: 1949, *Proc. Second Annual Reciprocal Meat Conference*, p. 117.

[27] Szczesniak, A.: 1963, *J. Food Sci.* **28**, 410.

[28] Taneya, S. and Sone, T.: 1962, *Ohyohbutsuri* (*Appl. Phys. Japan* in Japanese) **31**, 483.

[29] Taneya, S. and Sone, T.: 1962, *Ohyohbutsuri* (*Appl. Phys. Japan* in Japanese) **31**, 465.

[30] Taneya, S. and Sone, T.: 1962, *Ohyohbutsuri* (*Appl. Phys. Japan* in Japanese) **31**, 286.

CHAPTER 3

CONSISTENCY OF RESPECTIVE FOODS

3.1. Liquid Foods

The consistency of foodstuffs which are liquid near room temperature will be described in this section. Some typical liquid foods are soup, syrup, juice, puree, sauce, milk, etc. Although they are in the liquid state, fats or/and fibrous materials are suspended in the bulk of these systems, so that really they should be classified as emulsions or suspensions. The consistency of these liquid foods is much influenced by various factors such as the volume of the dispersed phase (i.e. concentration), interfacial state between the medium and the dispersed phase, shape of the dispersed phase, and so on, and many of the systems behave as non-Newtonian fluids.

3.1.1. MILK

Milk, which is a kind of secretion from the cow, can simply be considered as an emulsion of fat in an aqueous medium in which proteins are dispersed colloidally. The chemical composition of the components in milk is influenced by the growing conditions of each individual cow, i.e. climate and feed, and may produce differences in the consistency of milk.

The consistency of milk relates closely to the sensory term 'thickness'. As milk is an emulsion, the dispersed fat ascends to the upper part of the system during a period of storage, and the so-called cream line is formed. In the past, the concentration of fat in milk was estimated by the depth of the cream line. Now, the fat in milk is homogenized by means of a homogenizer in order to prevent the formation of the cream line. Casein and albumin are the main proteins in milk. The surface of the fat globules in milk is covered by a thin film, of approximately 5 mμ thickness which stabilizes the emulsion state.

The main component of the thin film is a complex of phospholipid and protein, and this lipoprotein acts as an energy barrier for stabilizing the system on the surface of the fat globules. The phospholipid in the lipoprotein consists mainly of lecithin and cephalin, and both components contain hydrophobic groups (stearic acid and oleic acid) and hydrophilic groups (choline, aminoethyl alcohol, pholic acid), so that the phospholipid behaves as an amphipathic molecule. Let us suppose that the densities of the phospholipid and the milk fat are 0.85 and 0.94, respectively, and that the molecular length of the phospholipid is 2.2 mμ, then the amount of monomolecular film of phospholipid adsorbed on the surface of the fat globules will be 0.38 g per 100 g of fat. The fatty acid groups of the phospholipid form a solid solution with triglyceride molecules, so it is possible to find a bi-refringent layer on the surface of the globules because of the orientation of the glyceride molecules. The thickness of the birefringent layer is about one-tenth of the diameter of the fat globules, and the thickness corresponds to a two-hundred molecules thick layer of the triglyceride. It should, however, be mentioned that the phospholipid combines loosely with the triglyceride molecules.

Although hydrophilic terminal groups in the phospholipid layer coordinate with the hydrophilic residues of some of the protein molecules in milk, other residues are ionized or form compact helixes because of many hydrophilic or hydrophobic residues in the protein molecules, so that the protein molecules generally form spheres. When the helix is dissolved the protein molecules become more flexible, and then the hydrophobic residues coordinate with the fat components while the hydrophilic residues coordinate with the aqueous phase. The main chains in the protein molecules then extend onto the surface of the fat globules with increasing surface pressure, and the surface viscosity of the globules increases substantially. If such flexible protein molecules adsorb onto the surface of the globules, three dimensional coagulation between the globules follows.

As has been illustrated by Jenness and Palmer [1], the protein molecules may form bi- or tri-molecular layers on the surface of the fat globules, and the thickness of such layers may range from 2.6 mμ to 3.8 mμ. These interfacial layers are mechanically so strong that they prevent coalescence between the fat globules, even when globules undergo creaming and drainage. Therefore, extraction of the fat components from the system should be made by chemical denaturation of the protein, or by using a strong mechanical shock.

As shown in Table 3.1, milk is more dense than water, and the specific heat of milk is lower than that of water, although the solid components occupy about 12% in milk. Because electrolytes are present, the electrical conductivity of milk is rather high, and the freezing point of milk appears at about 0.5 °C below that of water.

TABLE 3.1
Physical properties of milk

Density (20 °C)	1.0317 ~ 1.0336
Specific heat (20 °C)	0.95
Surface tension [dyn cm^{-1}]	46 ~ 47.5
Freezing point [°C]	−0.535 ~ −0.550
Electric conductivity [ʊ (ohm^{-1})] (20 °C)	0.004 ~ 0.0055

Einstein provided a relationship between the concentration (in volume fraction) of the dispersed particles and the viscosity of a dilute suspension using hydrodynamic theory for the case of non-interacting particles.

$$\eta = \eta_0 (1 + 2.5 \ \phi), \tag{1}$$

where η is the viscosity of the suspension, η_0 is the viscosity of the suspending fluid, and ϕ is the volume fraction of the dispersed particles. When the particles are fluid it is, however, necessary to consider the deformation of and internal flow within the dispersed particles. Therefore, at a temperature above the melting point of the fats, the viscosity of milk is given by [2]

$$\log \frac{\eta}{\eta_0} = 2.5 \frac{\eta' + 2\eta_0/3}{\eta' + \eta_0} \phi, \tag{2}$$

where η' is the viscosity of the fat components. Figure 3.1 shows the above relation, in which one can find a linear relationship between fat concentration and $\log (\eta/\eta_0)$ over a wide range of the concentrations.

Cox et al. [3] have illustrated, from the results obtained with many samples, that it is possible to find a functional relationship between the concentration of solid components and the viscosity of milk, but the viscosity is much influenced by the fat in the range of relatively high concentrations.

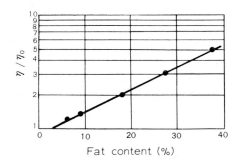

Fig. 3.1.   Plot of η/η_0 of milk against fat content.

In general, temperature affects the viscosity of liquid materials. Caffyn et al. [4] have observed a correlation between the absolute viscosity of milk and the temperature, as shown in Figure 3.2, i.e. the viscosity decreases rapidly with increasing temperature up to about 40 °C, and it decreases slowly with increasing temperature above 40 °C. The dependence of the viscosity of milk upon temperature can be given by

$$\eta = Ae^{B/T} \tag{3}$$

which corresponds to Andrade's equation. From the standpoint of the theory of the absolute reaction rate proposed by Eyring [5], the constant B in Equation (3) is equivalent to E/R where E is the activation energy of flow. Figure 3.3 shows a plot of log viscosity against the reciprocal of absolute temperature, and one can identify the change of slope in the linear relationship at about 40 °C. This phenomenon may be brought about by melting of the rigid fat globules around this temperature.

Although it is possible to consider diluted milk as a Newtonian fluid, Poiseuille's law cannot be applied to the concentrated sample;

$$\eta = \frac{1}{4}\left(\frac{PR/2L}{V/\pi R^3}\right),$$ (4)

where P is the pressure, R is the radius of the capillary tube, L is the length of the capillary tube, and V is the volume of sample which flows through the capillary tube per unit time. In Equation (4), the term $PR/2L$ represents shearing stress, and $V/\pi R^3$ equals rate of shear, so that a linear relationship between these two terms cannot be expected for concentrated milk. One may find five patterns of flow curve for concentrated milk in plots of $V/\pi R^3$ against $PR/2L$ such as α, β_1, γ, δ, β_2, as shown in

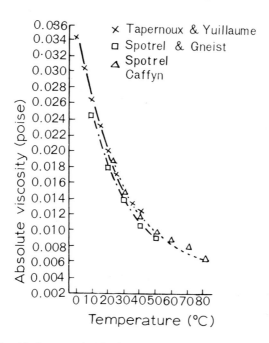

Fig. 3.2.   Relationship between the absolute viscosity of milk and the temperature.

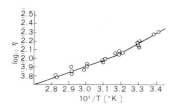

Fig. 3.3.   Dependence of milk viscosity upon temperature.

Figure 3.4. An interesting point in these patterns is the dependence of the slope of the curve upon capillary size, as seen in γ type and δ type patterns; this phenomenon is called the sigma effect, and it may be caused by the heterogeneous flow of dispersed particles along the capillary tube, i.e. the particles tend to collect at the centre of the tube when using thinner capillary tubes, so that the viscosity of the dispersed system appears to be low.

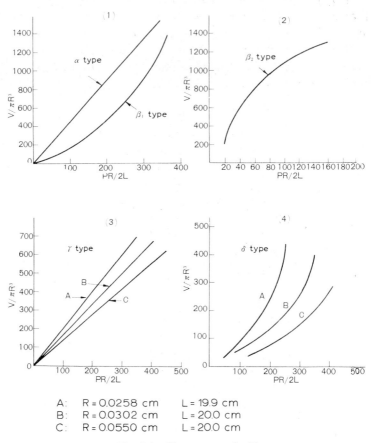

A:	R = 0.0258 cm	L = 19.9 cm
B:	R = 0.0302 cm	L = 20.0 cm
C:	R = 0.0550 cm	L = 20.0 cm

Fig. 3.4.   Flow curves of milk.

As has been described already in this section, it is possible to prepare well defined small globules by homogenizing the milk. Cox *et al.* [3] have measured the viscosity of homogenized milk in a temperature range of 20 °C–80 °C, and they have derived a relation for obtaining the viscosity of the homogenized milk η at any temperature from the viscosity at a standard temperature η_θ, as follows:

$$\eta = \frac{\eta_\theta}{1 + \alpha\theta + \beta\theta^2}.$$
(5)

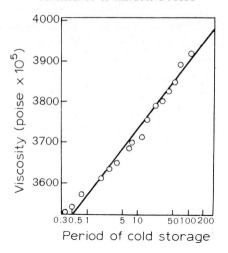

Fig. 3.5.   Viscosity change of homogenized milk after pasteurization at low temperature.

Pasteurization and cooling procedure clearly increase the viscosity of homogenized milk; this may be due to hydration of the denatured proteins, and by change in the amount of adsorbed protein molecules on the surface of the fat globules. Whitnah et al. [6] have provided experimental data (Figure 3.5) on the relationship between viscosity and the aging time of homogenized milk, which was prepared from milk with a fat concentration of 4±0.05%, and with a solids content of 13.0±0.1%, which was homogenized by using a pressure of 2500 lbs. in.$^{-2}$ at 59 °C. They have also proposed an empirical equation from the above results, i.e.

$$\eta = a + b \,(\log \text{ aging period}) \tag{6}$$

Sometime before the above investigation, Bateman and Sharp [7] pointed out that homogenization produces an increase in the viscosity of milk, and that pasteurization lowers the viscosity of milk slightly.

In the homogenizing process, the fat globules in milk become smaller as the homogenizer pressure increases, as shown in Table 3.2.

TABLE 3.2
Change of fat globule size in milk due to the homogenization*

Pressure (lbs/in.²)	Mean diameter of fat globules (µ)					
	0	500	1500	2500	3750	4500
Buttenberg (1903)					0.80	
Wiegner (1914)	2.9				0.27	
Marquardt (1927)	3.88			1.56		0.97
Tront et al. (1935)	3.89	2.50	1.91	1.38		0.97
Tracy (1948)	3.71	2.39	1.40	0.99		0.97

* Trout, G. M., Homogenized Milk, Michigan State Univ. Press, East Lansing (1950).

If 5% of the fat globules in milk with an average diameter of 5μ are homogenized and reduced in size to $1\,\mu$ diameter, the total surface area of the globules increases from $720\,000\ \mathrm{cm^2/1}$ to $3\,456\,000\ \mathrm{cm^2/1}$, so that the thickness of the adsorbed protein layer on the surface of fat globules becomes thinner, and casein protein in the medium may be adsorbed on to the surface of the globules.

Puri *et al.* [8] have measured the viscosity of 283 milk samples prepared from bisons, cows, and goats, and they have shown that the viscosity increases with increasing concentration of the solid components, and that the viscosity of milk obtained from goat is more temperature sensitive than that of cow's milk. They have also studied the effect of additives on the viscosity of milk; 1% of sucrose or starch has no effect, but the same amount of gelatine produces an increase in the viscosity of milk. Dehydration, which occurs in milk when ethyl alcohol is added, produces an increase of the viscosity according to the concentration of ethyl alcohol which is used, and ultimately the system coagulates. In the case of a fixed concentration of the solid component, the slope of the correlation curve between the viscosity of milk and the amount of added alcohol increases as the concentration of fats in milk increases, as shown in Figure 3.6.

The major component of milk is protein, of which casein occupies a large part. Casein molecules disperse colloidally in the medium as micelles, so that milk has an opaque appearance. The structure of the casein micelle is built up by the coordination of many molecules through calcium bridge and dipole-dipole interactions between the molecules. Additives such as alkali, urea, calcium salts, or polyphosphoric acid salts tend to reduce the micelles to the individual molecules by dissolving the calcium bridges and hydrogen bonds.

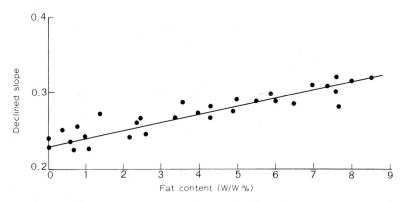

Fig. 3.6. Plot of declined slope on the viscosity change of milk against fat concentration obtained by adding alcohol to the sample.

When calcium hydroxide is added, the viscosity of skimmed milk increases rapidly within a short period of time, then it decreases slowly for a few days, and finally it increases again to form a gel, as shown in Figure 3.7. Urea also plays a part in the viscosity changes described above. Viscosity changes in milk are due to changes in structure of the casein micelles; the first viscosity increase relates to swelling of the micelles together with increased interaction between the micelles. A decrease in the viscosity at the next stage corresponds to the disintegration of these swollen micelles into individual molecules. The maximum value of viscosity reached in the

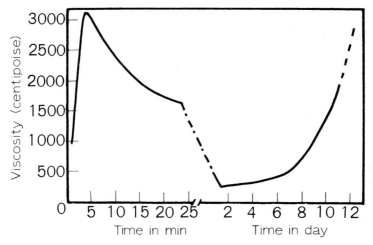

Fig. 3.7.   Viscosity change of condensed skim milk due to the presence of calcium at 4°C
(milk solid: 22.7%, pH: 11.7).

first stage increases proportionately with increasing temperature, but it increases inversely with temperature when hexametaphosphate is used as an additive, i.e. in the latter case one may obtain maximum viscosity when hexametaphosphate is introduced into the system at a low temperature.

3.1.2. SYRUPS

The term syrup is used generally for millet jelly, honey, theriac, aqueous solutions of sucrose, and maple syrup.

Honey is an aqueous concentrated solution of fructose and glucose, and it also contains a little amount of carbohydrate, protein, minerals, vitamins, and so on. Honey generally shows Newtonian flow, and the value of the viscosity depends on the concentration of sugar components at a fixed temperature. Some honeys, however, show very non-Newtonian flow. A sample which is obtained from the moorland heaths in England, behaves as a thixotropic body. Pryce-Jones has demonstrated that this honey can be identified as a false body, i.e. the honey becomes fluid with shear, and it changes to a solid when agitation ceases, so that a sol-gel inversion of the sample occurs with shear. The thixotropic characteristics of Heath honey are caused by the presence of a small amount of protein, so that the sample is gelled by heating around 65°C.

As shown in Table 3.3, fructose and glucose form the main components of honey, the moisture content ranges from 10% to 20%, while sucrose is a very minor com-

TABLE 3.3

Composition of various honeys

Component	South Carolina (cotton flower)	California	Florida	New Zealand (clover)	England (heath)	South Africa (euca- lyptus)	Switzerland (nectar)
Moisture	10.2	17.4	18.4	16.6	21.2	18.2	19.3
Fructose (L)	44.6	35.7	47.0	39.7	38.0	36.2	31.3
Glucose (D)	39.8	40.1	24.2	38.4	35.7	35.6	24.6
Sucrose	1.8	3.2	4.6	1.9	1.2	1.2	2.2
Melezitose	–	–	–	–	–	–	8.7
Dextrin	–	–	–	–	0.2	0.1	11.3
Protein	–	–	–	–	1.8	0.2	0.3
Dextran	–	–	–	–	–	5.3	–
Others	3.6	3.4	5.8	3.4	1.9	3.2	2.3
Ratio of D/L	1.12	0.88	1.95	1.03	1.06	1.02	1.27

ponent. Therefore, the concentrations of fructose and glucose play a big part in the consistency of honey.

Some kinds of honeys contain dextran which causes an abnormal viscosity (dilatancy) of honey.

As far as Newtonian honey is concerned, its viscosity is very much influenced by the moisture in the system. For example, a honey whose moisture content is 14.5%, has a viscosity of 485 poise, while the viscosity becomes only 28 poise when the moisture content rises to 21.2% (i.e. the viscosity falls to 1/20th of the former case). Because of the regular dependance of the viscosity of honey on temperature, one can obtain a relationship between log viscosity and the reciprocal of absolute temperature, and it is possible to apply Andrade's equation to the data. Also one can find a linear relationship between log viscosity and the reciprocal of the moisture content, as shown in Figure 3.8, so that a general equation for expressing a correlation between the viscosity of honey, its moisture content, and the temperature can be derived by [9]

$$W = (62\,500 - 156.7\ T)\left[T(\log \eta_T + 1) - 2287(313 - T)\right], \qquad (7)$$

where W is the moisture content, T is the absolute temperature, and η_T is the viscosity at temperature T.

On the other hand, Munro [10] has plotted the viscosity of various honeys against temperature, as shown in Figure 3.9, and derived a non-linear relationship. The pronouncedly non-linear curve E–E corresponds to the honey from heather which contains protein, and is described as a thixotropic honey.

Pryce-Jone [11] has tried to obtain flow curves of honey from heather using a Couette type viscometer. Figure 3.10 shows the results, in which the viscosity decreases from 460 poise (A in Figure 3.10) to 70 poise (B in Figure 3.10) with slowly increasing rate of shear from 50 s^{-1} to 900 s^{-1}, then the viscosity increases again to 230 poise (C in Figure 3.10) when the rate of shear decreases again from 900 s^{-1} to 50 s^{-1}. However, one finds that it is not possible to superpose the curve A–B on to the curve *B–C* in Figure 3.10. This phenomenon is clearly defined as thixotropy, i.e. it takes a period of time to recover the original level of viscosity.

Dependence of viscosity upon rate of shear surely implies that the fluid is not Newtonian. The other thixotropic honey is Manuka (Leptospermum Scoparium) which is prepared in New Zealand. This honey also contains about 1% of protein like Heath honey. It is difficult to discriminate Manuka from Heath because of the very similar sugar crystal forms in both honeys, although Manuka can be characterized by the pollen.

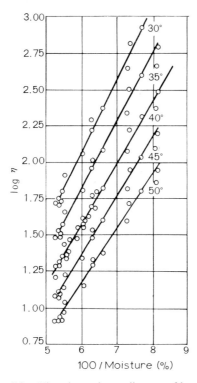

Fig. 3.8.   Viscosity-moisture diagram of honey.

Pryce-Jones [12] has pointed out that some honeys show dilatant flow which means that the viscosity increases with increasing shear. The honey 'Eucalyptus ficifolia' exhibits dilatancy. Dilatant honey can be characterized by the 'spinnability' phenomenon.

Figure 3.11 shows the correlation between rate of shear and the apparent viscosity of Eucalyptus honey and one finds a rapid increase in the apparent viscosity with increasing rate of shear. Dilatancy of Eucalyptus honey is brought about by the presence of 7.2% of dextran $[(C_6H_{10}O_5)_n, n=8000]$ which may be synthesized microbiologically by Leuconostoc mesenteroides. If one fractionates out this dextran from the honey the dilatancy of the system should disappear, and the system becomes a Newtonian fluid. On the other hand, if one introduces a small amount of dextran

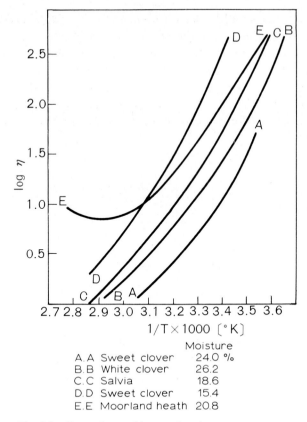

		Moisture
A.A	Sweet clover	24.0 %
B.B	White clover	26.2
C.C	Salvia	18.6
D.D	Sweet clover	15.4
E.E	Moorland heath	20.8

Fig. 3.9.   Dependence of honey viscosity upon temperature.

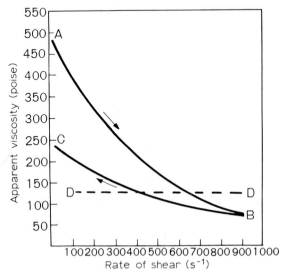

Fig. 3.10.   Hysteresis curve of honey.

into Clover honey which is a Newtonian fluid, the system shows dilatancy. These observations indicate that dextran is the cause of dilatancy in honey.

Dilatant honey is not only an abnormal fluid but it is also an elastic body. If a rod is immersed in dilatant honey, and it is rotated, the honey crawls up the rod due to the effect of normal stress. This phenomenon is known as the Weissenberg effect after the discoverer.

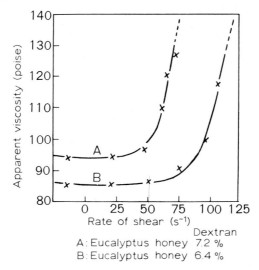

Dextran
A: Eucalyptus honey 7.2 %
B: Eucalyptus honey 6.4 %

Fig. 3.11.   Dilatancy of honey.

Theriac and syrup, which are made by mixing sucrose with maple syrup, behave as Newtonian fluids, and their viscosity depends upon the concentration of sucrose. The log viscosity can be plotted linearly against the concentration of sucrose. The National Bureau of Standards (U.S.A.) has provided full details of the viscosity of aqueous sucrose solutions, and these are widely utilized for reference purposes. Figure 3.12 shows the correlation between the viscosity of aqueous sucrose solutions and temperature at various concentrations of sucrose.

Maple syrup is standardized by its water content up to 35%, and contains solid components which consist of a large amount of sucrose and a small amount of reducing sugar. Maple syrup is a Newtonian fluid, because the sugar components generally do not induce any abnormal flow into the system.

Sugar syrup is prepared from a diluted aqueous sugar solution by concentration in vacuum, so that, when there is a relatively low pH value in the system, the amount of reducing sugar increases during the procedure with increasing sugar concentration. The viscosity of maple syrup is lowered by increasing the concentration of reducing sugar in the system, as shown in Table 3.4.

The viscosity of aqueous sugar solution is also much influenced by the presence of inorganic salts. For example, sodium salts and potassium salts effectively lower the viscosity, while calcium salts increase the viscosity.

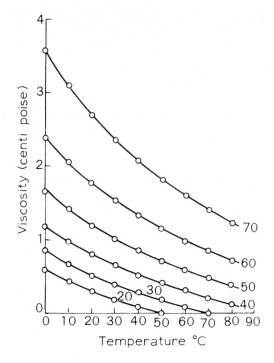

Fig. 3.12.   Relationship between the viscosity of aqueous sucrose solutions and the temperature.

Millet jelly is prepared by hydrolysis of starch using acids or enzyme amylase, so that the main components of millet jelly are glucose, oligosaccharide and polysaccharide. This system generally shows Newtonian flow.

3.1.3. JUICE AND PUREE PREPARED FROM FRUITS AND VEGETABLES

Juice and puree extracted from fruits or vegetables contain so many dispersed materials such as insoluble particles, fibrils and colloidal polymers that most of the systems behave as non-Newtonian fluids. Harper [14] has shown that concentrated purees appear to be pseudo-plastic.

TABLE 3.4

Dependence of viscosity of aqueous sugar solution upon invert sugar content [13]

% of invert sugar and sucrose in solid component		Viscosity of aqueous 65% sugar solution at 22°C
Invert sugar	Sucrose	
0	100	133
10	90	127
50	50	97

From the view point of sensory assessment of juice quality, the important thing is not only freshness of taste and flavour but the sensation produced on the tongue. Also the fluidity, which may be identified on transferring juice from the tin to a glass, is important for assessing the consistency.

Juice is made by a process which involves crushing and pressing juice out from fruits or vegetables, and pasteurization, so that the various tissues of the raw materials are more or less mixed into juice. Such mixing of tissues affects the consistency of the system. Figure 3.13 shows a cross section of a tomato, which roughly illustrates the four main parts, i.e. (A) shell tissue, (B) centre tissue, (C) free juice near the seeds, and (D) seed envelope. The shell tissue, which is covered by the skin, occupies about 48% of the tomato volume, but the seed envelope occupies only 8% of the volume. The tissue cells are nearly spherical, and the interior of each cell is filled with protoplasm. Cellulose forms a network structure at the cell wall together with pectin, so contributing mechanical strength to the cell.

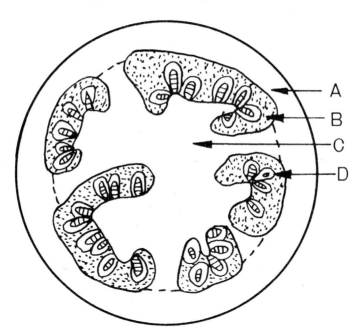

Fig. 3.13.   Cross section of tomato.

Whittenberger and Nutting [15] have prepared tomato juice at 85 °C, and they measured the viscosity of mixed samples of juice with tissue using a Brookfield's viscometer. They also measured the viscosity of protoplasm, which had been extracted from the mixed juice, using an Ostwald's type viscometer. The results are shown in Table 3.5.

If the viscosity of the juice from one part of the tomato is compared with the viscosity of juice from the other parts (Table 3.5), one can find that the viscosity has a

TABLE 3.5

Consistency of tomato juice

Composition	Ratio to whole tomato (%)	Apparent viscosity of juice (cp)		Viscosity of serum (cp)
		Original	Homogenized	
(A) Shell tissue	48	155	405	2.6
(B) Centre tissue	26	145	335	2.4
(C) Free juice	18	11	15	1.3
(D) Seed envelope	8	432	700	25.6

maximal value in the seed envelope juice, and that it is at a minimum in the free juice near the seeds. The viscosity of shell tissue juice is quite similar to that of centre tissue juice. When one homogenizes the mixed juice from each part, the viscosity of the juice from every part increases pronouncedly, although the ranking order of each viscosity value is not influenced by the homogenization procedure. Increase of the viscosity may be due to the fibrous materials dispersing into the system from the cell wall following rupture of the cells through the homogenization process. The viscosity of protoplasm extracted from the shell tissue, centre tissue, and free juice, re-spectively, has a low value, but protoplasm extracted from the seed envelope is very viscous and slightly elastic because of the presence of a large amount of pectin.

The viscosity of tomato juice is also much influenced by the level of maturity of the tissue. It is, however, so difficult to provide a standard for the grade of maturity that heat treatment is applied to soften the tissue, to make it easy to separate the tissues from each other, and to destroy any enzymatic action on the pectin materials. Hand et al. [16] have suggested that a preheat treatment is necessary for preparing tomato juice, because this process increases the viscosity of the juice, and also because high temperature affects the viscosity of protoplasm.

Whittenberger et al. have reported that the electrolytes which are contained in tomato, or are added to the juice, are responsible for keeping the viscosity of juice at a relatively low value. Therefore, it is possible to thicken the juice by fractionating out the soluble pectin, organic acids and inorganic salts from the juice. The cell wall occupies about 6% of the solid components in tomato; this value corresponds to half of the total insoluble material in tomato. With decreasing concentration of electrolytes in the juice the cell walls begin to swell, so that non-electrolytes such as sucrose, glycerin, ethyl-alcohol, etc., are useful for preserving the viscous state in juice.

Although orange juice is generally a diluted suspension of the solid components, the suspended pulps play a big part in the consistency of the juice when it is in the concentrated state. Viscous orange juice is not desirable from the viewpoint of the user or manufacturer because it is difficult to remove from a tin, and also because cooling after pasteurization would not be effective because of the low thermal con-ductivity of the concentrated system.

Ingram [17] has shown that concentrated orange juice is more viscous than aqueous sucrose solution containing the same sugar concentration because of the pectin in the juice, and that the viscosity of orange juice increases greatly at 65% concentration of the solid components. Ezell [18] has observed that the viscosity of juice increases with increasing concentration of the pulp at a fixed sugar concentration. It seems that the pectin elutes into the bulk phase from the pulp, and that the pulp itself behaves as a solid dispersed phase in the suspension.

Flow curves of suspensions can be represented by Casson's equation, i.e.

$$\sqrt{f} = \sqrt{s} + \sqrt{\dot{\gamma}} \tag{8}$$

where f is the yield value, s is the shearing stress, and $\dot{\gamma}$ is the rate of shear.

The viscosity of concentrated orange juice has been measured by Charm [19] at $0\,°C$ using a rotational viscometer. When using a rotational viscometer, the torque T which acts on the rotational cylinder per unit length corresponds to the shearing stress s, and the rotational speed of the cylinder N relates to the rate of shear $\dot{\gamma}$. Then, if Casson's equation is applied to the orange juice, $\sqrt{T}$ should be plotted linearly against $\sqrt{N}$. Figure 3.14 shows that Casson's relation is satisfied by some fruit juices. It must be mentioned, however, that Casson's relation is only a phenomenological representation of the flow curve, so that it may be not possible to identify the cause of gelation of the pectin materials, or the contribution of pulp to the non-Newtonian properties of juice.

In order to clarify the above problem, Berk [20] has tried to obtain information on the relationship between pulp size and the viscosity of concentrated orange juice, using a technique in which the concentrated juice has been irradiated ultrasonically by 20000 Hz (amplitude: 10 μ) and the change of shape in the pulp has been observed microscopically. As shown in Figure 3.15, the viscosity of the irradiated orange juice decreases pronouncedly with increasing irradiation ultrasonic energy. Berk has also pointed out that this lowered viscosity is not changed by aging over 30 days. From the results of microscopic analysis it appeared that the decrease of viscosity is brought about by a change in the shape of the pulp from a film state to a fibrous form. This study may help the technology of juice processing, i.e. it is possible to obtain stable concentrated juice by resolving the problem of the thermal conductivity of juice in the thermal treatment.

TABLE 3.6

Flow properties of tomato paste in pipe

Volume of flow Q (cm^3 s^{-1})	Pressure P (dyn cm^{-2})	Pressure loss P/L (dyn cm^{-3})
0.1	183000	1500
0.5	275000	2250
1.3	348000	2850
4.3	438000	3900
9.9	622000	5100

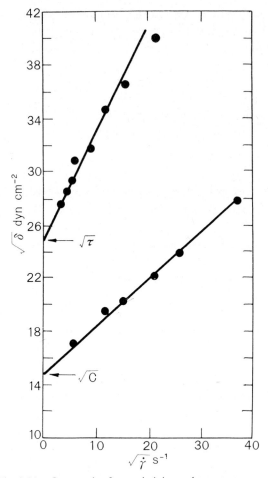

Fig. 3.14.   Casson plot for apple juice and tomato puree.

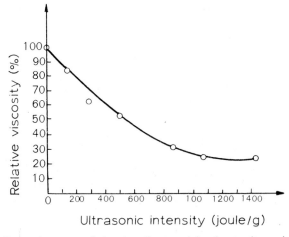

Fig. 3.15.   Decreasing curve of viscosity of orange juice due to ultrasonic treatment.

Although purees and pastes are generally non-Newtonian fluids, the flow properties are very much influenced by the manufacturing process applied to the fruits. Table 3.6 shows the correlation between the pressure loss and the rate of flow of tomato paste in a tube viscometer. If one tries to obtain the flow constant of the tomato paste as a fluid obeying the power law, using Charm's method, one finds that the apparent viscosity of the sample alters pronouncedly with the power index when applying a fixed pressure loss or rate of flow. This suggests that it is necessary to consider the structural viscosity of paste foods when transporting the system in the manufacturing process. As far as banana puree is concerned, Charm [21] has already measured the relationship between the shearing stress and the rate of shear at various temperatures using Merrill's viscometer (Table 3.7). Smit and Nortze [22] have

TABLE 3.7
Flow properties of banana puree (shearing stress ~ rate of shear)

Rate of shear $(s^{-1}) \times 10^3$	1	1.5	2	3	4	5	6	7	Temp. (F)	b dyn s^3 cm^{-2}	S
Shearing stress $(dyn\ cm^{-2}) \times 10^{-3}$	1.59	1.95	2.18	2.65	3.05	3.05	3.75	3.97	68	68.9	0.460
	1.32	1.56	1.80	2.16	2.50	2.85	3.05	3.30	77	57.2	0.407
	1.25	1.50	1.75	2.10	2.36	2.70	2.96	3.10	107	52.6	0.486
	1.06	1.22	1.37	1.62	1.80	2.01	2.10	2.21	120	41.5	0.478

tried to obtain a correlation between the concentration and the fluidity of tomato juice samples prepared from San Marzano tomato and Pearson tomato (Figure 3.16). Tomato paste is usually passed through a Finisha sieve in the manufacturing process, and the consistency of the paste is reduced to a greater extent by passing through a fine sieve than it is by treatment with a coarse sieve. The consistency increases greatly, on the other hand, when one takes a long period of time for processing the crushed pulp and preheating. This fact disagrees with the belief that the consistency

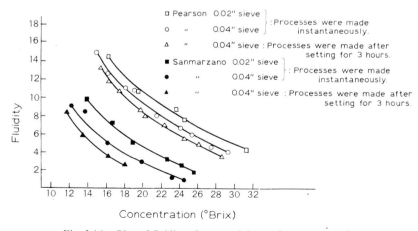

Fig. 3.16. Plot of fluidity of tomato juice against concentration.

is lowered by the preheating process because pectin materials are decomposed by the active pectinase near the skin. Actually, the amount of calcium pectinate decreases with increasing time for the preheat treatment, while the high consistency of paste may be brought about by the formation of secondary structures within the dispersed phase. Paste prepared from San Marzano tomato shows a higher consistency than the paste prepared from Pearson; this may be due to the fact that San Marzano tomato contains a large amount of pectin.

3.2. Consistency of Gel-Like Foods

Most processed foods and raw foods can be regarded as gel-like materials from the view point of their physical properties. Also most of the raw materials obtained from animals and vegetables contain some substances which cause gel formation in the systems. These substances are mainly polymerized carbohydrates such as starch, agar, etc., or proteins, and they confer a gel state on the dispersion in the aqueous medium. We have many gel-like foods such as puddings, yokan (sweet jelly of beans), jelly, tôfu (bean curd), cottage cheese, cheese, cooked egg white, and so on, which contain polymerized carbohydrates or proteins.

Such gels are able to preserve the gel state even in very diluted systems. They have a high elasticity which helps to retain the shape of the foods. For example, gels prepared from agar, pectin, starch, etc., exhibit on elasticity of 10^7 dyn cm^{-2} or less, which is a rather low value in comparison with that of other solid materials, but these gels deform easily and have a high extension ratio. Gels can generally be divided into various categories from the characteristics of the system e.g. swollen gel (polymer molecules swell in the medium), thermo-reversible gel (system reversibly shows sol-gel inversion due to temperature change), and thermo-irreversible gel (gelation is brought by heating the system). The elasticity of a gel is denoted as entropy elasticity because induced stress in the system increases with increasing temperature, while the elastic deformation of crystals or other solid materials relates to the increase of internal energy (energy elasticity).

Formation of a gel is brought about by interaction between the dispersed molecules, including friction due to thermal fluctuation of the molecules. This is caused by swelling of the molecules due to solvation in the medium. The molecules expand in the medium, and bind partially with each other, so that a network structure is formed. Also long-range interaction forces between the molecules or particles contribute to the gel state.

In the case of aqueous media, the amount of water taken into the gel is quite changeable. Excess water in the system brings about an inversion from the gel state to a sol due to dilution of the swollen molecules. The reduced viscosity of a diluted aqueous agar solution increases with decreasing concentration of agar in a range of extremely low concentrations, as shown in Figure 3.17. If one presses a diluted gel such as this, the bound water in the network structure is removed. Free water oozes out spontaneously from the surface of the gel system. This phenomenon, denoted as syneresis, appears with increasing number of linkages in the network, because the bound water may be squeezed from the system by the tightening structure of the network. When an electrolyte is introduced into the system the dispersed polymer molecules also shrink due to the neutralization of electric charges on the stretched chains of polymer molecules by the ions provided by the electrolyte. For example, calcium ions strongly check the dissociation of polymer molecules, so that the increase of viscosity previously exhibited at low polymer concentrations is no longer shown.

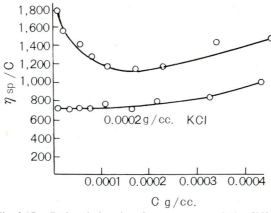

Fig. 3.17.   Reduced viscosity of aqueous agar solution [23].

3.2.1. STARCH GEL

Starch is a polymer of glucose; 1,4 linkages between glucose molecules forms amylose as a linear polymer, and a mixture of 1,4 and 1,6 linkages builds up amylopectin which possesses side chains from amylose molecule. Figure 3.18 shows their structures.

In the starch granule, the above two kinds of molecules are packed so regularly and tightly that the starch granule appears birefringent in a polarizing microscope.

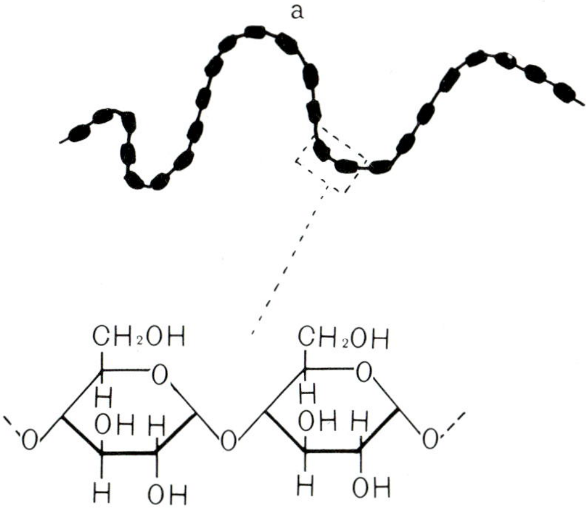

Fig. 3.18a.

b

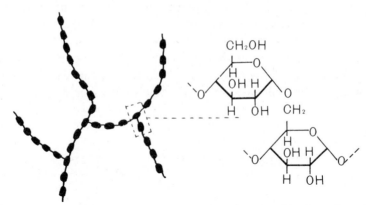

Fig. 3.18a–b.   Chemical structure and configuration of amylose (a) and amylopectin (b).

Starch granules are insoluble in cold water, but swell with water when heated, and then the birefringent layer disappears from the granules. Starch granules are still spherical in the initial stage of swelling. As the temperature increases molecules of amylose and amylopectin are gradually isolated, and eventually the granules disappear from the system. The chemical properties of starch, such as an affinity to iodine, vary according to the vegetable origin (Table 3.8).

The intrinsic viscosity of aqueous starch solution depends upon the specific volume and degree of polymerization of the starch molecules. The shape of the starch molecule in the fluid medium is much influenced by the condition and state of the system. Linear chains align themselves with other molecules in the medium, during cooling, due to the formation of hydrogen bonds, and this procedure results in the formation of micellar structures. Such micelles are insoluble in the medium (Figure 3.19).

The molecular weight of amylopectin is about 10^6, which is larger than that of amylose. The length of the side chain in amylopectin corresponds to 19–20 units of glucose. Solutions of amylopectin are so stable that it is difficult to have an orientated conformation of the molecules. The intrinsic viscosity of such solutions is in the range of 120–190.

In general, the native starch in raw vegetables is used directly for preparing foods, so that the viscosity of aqueous solutions of starch is an important characteristic of the systems. Schoch has tried to apply the Brabender Amylograph for testing the consistency of starch. The starch sample is kept at 95 °C for one hour after heating, and then it is maintained at 50 °C for one hour after cooling.

Viscosity changes in the sample during this treatment are shown in Figure 3.20, from which it follows that the flow properties depend on the origin of the raw material. Tapioca starch granules swell rapidly in the medium when it is heated, and the viscosity reaches a maximal value within a very short time, but this high viscosity soon falls to a low value at the same temperature. Potato starch exhibits similar behaviour. Cereal starches, however, show a relatively low degree of swelling and

TABLE 3.8

Properties of various starches [24]

Kind of starch	Shape of granule	Granule size (μ)	Affinity with iodine	Amylose content (%)	Intrinsic viscosity	Degree of polymerization
Seaweed (Florida)	Spherical	15	0.1	1		
Rice family						
Amylomaize	Spherical	25	9.9	52	180	1300
Barley	Spherical	20	4.3	22	250	1850
Oat	Spherical	25	5.1	27	180	1300
Wheat	Spherical	30	5.0	26	280	2100
Zeamays	Spherical	30	5.5	28	150	1100
Bean family						
Broad bean	Oval	30	4.5	24	240	1800
Pea (Smooth-seeded)	Oval	30	6.6	35	180	1300
Pea (Wrinkled-seeded)	Mixed type	40	12.5	66	140	1000
Subterranean stem						
'Ichihatsu'	Oval	30	5.0	26	240	1800
America 'Bofu'	Spherical	15	2.1	11	590	4400
Potato	Oval	40	4.3	23	410	3000
Seed and fruit						
Apple	Spherical	10	3.6	19	200	1500
Banana	Oval	35	3.0	16	240	1300
Mango	Oval	25	4.5	24	240	1800
	Spherical	5	3.7	19	220	1600
	Spherical	5	4.2	22	200	1500
Soft type						
Maize	Spherical	15	0.1	< 1		

consistency in the initial stage of heating. Then the consistency increases with decreasing temperature, which means that the starch molecules tend to orientate with each other. Linked Solgum starch does not show a maximal consistency, and also it is not easy to observe dissolution of the granules. It may follow from consistency data obtained with the Brabender Amylograph that the starch which contains a large amount of amylopectin does not form a hard gel but a viscous paste, that corn starch forms a stiff and brittle gel, that wheat starch gives a softer gel than corn starch gel at equivalent concentrations, and that the starch prepared from oats or rice provides a soft and fluid gel.

The consistency of uncooked foods is not influenced by the presence of starch, although starch granules play a small part in the plasticity of the system, because most of the starch granules have little affinity to the water. In the case of potato or rice, the important factor which affects the consistency of cooked foods is not only the amount of free starch isolated from the granule but also the quality of the starch.

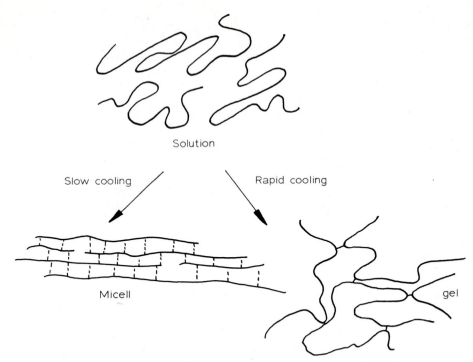

Fig. 3.19.   Schematic patterns of configuration change of amylose molecule by thermal treatment.
Parameter: weight of starch/500 ml of system.

An excess of isolated free starch confers an undesirable viscous state on the foods.
    Starch consistency is much influenced by the kind of raw vegetable, from which it
is derived, and this should be taken into account when preparing foods. It is possible

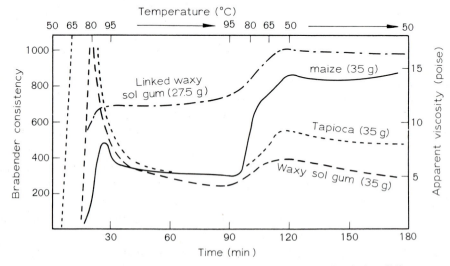

Fig. 3.20.   Brabender consistency curve of concentrated starch solutions [25].

to utilize the improved starch which is obtained from native starch by reduction with enzyme, acid, or alkali. Figure 3.21 illustrates the amount of water required to achieve the same consistency with various starches, including the improved starch.

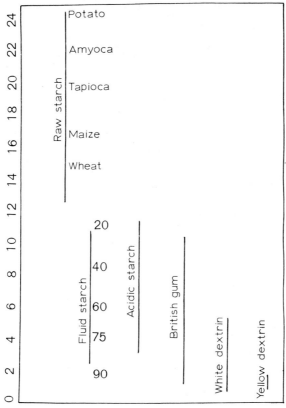

Fig. 3.21.   Range of necessary water contents for preparing starch solutions of similar consistency [26].

   Alpha (α) starch is prepared by quickly drying heat treated starch using a roll dryer, so that the α starch can immediately be swollen and gelled when water is added. This starch is useful for making puddings and pies.

   Acid treated starch, which is prepared by decomposition of pasty starch by diluted sulfuric acid or hydrochloric acid at 50 °C, consists of decomposed polysaccharide molecules. This starch is used to obtain a fluid paste by heating, and to provide a rigid gel at relatively low temperature. Soft candy, jelly beans or gel-state cakes are made with acid treated starch.

3.2.2. PECTIN GEL

The pectin molecule consists of a linear chain of galacturonic acids whose carboxyl groups are partially esterified with methyl alcohol, as follows;

Although the molecular weight of pectin is about 200000, pectin actually appears together with other polysaccharides such as arabans or galactans as a complex, so that pectin is apparently located in a much larger molecule. Fruits and vegetables contain many pectin complexes, which are concentrated around the cell wall or near the skin, because the pectin complex plays an important part in the framework and shape of fruits and vegetables. An extracted and purified pectin swells and disperses in water, and the system becomes a viscous solution, from which one can prepare a gel with sugar or acid. Therefore, pectin is utilized for making desserts such as fruit jellies or jams.

In 1947, a test method for pectin jelly was proposed by the National Preservers Association in the U.S.A., because the manufacturers were faced with the necessity to provide a standard for testing jellies due to the increased demand. Next year, IFT organized a research group in order to discuss methods and apparatus for providing a standard of jelly [27]. Accordingly, a meeting was held, and this was attended by six pectin manufacturers, four consumers and three government scientists. The following conclusions on test methods for jelly were reached. Pectin is a hydrophilic colloid in an acidic aqueous medium.

The system forms a gel due to the network structure of dispersed pectin molecules over a range of hydrogen ion concentrations, so the test sample should have a solid content of $7.05 \pm 0.5\%$ in water at a pH of 3.10 ± 0.05.

However, the manufacturers use very many different techniques for processing pectin. For example, one manufacturer mixes dried pectin with other materials, while another uses pectin in the solution state. Some manufacturers treat pectin in a vacuum vessel at a low temperature, others heat the sample directly, pectin is treated in hard water by some manufacturers or it is dissolved in soft water, and there is no standardized procedure as to how to add acid or how to heat the sample. Techniques for mixing sweetening materials with pectin also vary with different manufacturers. Therefore, it is not possible to provide a standardised testing method out of all the various procedures described above.

Accordingly, the IFT committee established a standard method 5–54 (May, 1954) for evaluating pectin materials. Part of the methods and apparatus has been described in the previous chapter.

Olsen [28] and Olsen et al. [29] have indicated that the following points should be considered regarding gel formation by pectin with acid and sugar:

(1) In solution pectin is a hydrophilic colloid with negative charges.

(2) Sugar acts as a dehydrating agent for pectin molecules.

(3) Hydrogen ions play a big part in the formation of a network structure by pectin, as the hydrogen ion makes decreases the negative charges on the pectin molecules.

(4) It takes a long time to obtain dehydration equilibrium in the pectin system. The rate of dehydration is influenced by the kind of pectin and by temperature.

(5) Both rate of dehydration and rate of sedimentation increase with increasing concentration of hydrogen ions.

(6) Maximum jelly strength leads to an equilibrium.

The stiffness of pectin gel is mainly brought about by dipole-dipole interaction between the molecules, even though Van der Waals' forces between the molecules, or the ionic strength, are reduced.

Cheftel and Mocquard [30] have studied the rheological properties of a concentrated gel of sugar-methoxy pectin mixture, and they have reported that the sample does not exhibit any internal strain in the large deformation region, although the sample does show typical plastic flow under a small stress. On the other hand, Owens et al. [31] have illustrated that the mixed gel of sugar and methoxy pectin behaves like a perfectly Hookean body in the range of small deformations. Actually, it is necessary to have a setting time for obtaining pectin jelly with a fixed jelly strength, although the setting time is influenced by temperature. Doesburg and Grevers [32] have mentioned that factors affecting the setting time include the ash content, degree of esterification of the molecules, and pH value, i.e. the setting time of a pectin gel which contains a small amount of ash increases with decreasing esterification grade, but the setting time becomes shorter in the low pH range. The presence of sugar in the sample decreases the setting time.

When the amount of methyl ester in the pectin molecules is below 7%, one may experience difficulty in preparing gels using sugar and acid, although the presence of calcium promotes gelation of the system. Calcium bridges may be formed between the carboxyl groups in the pectin molecule by the ionization of calcium.

3.2.3. VEGETABLE GUMS AND VISCOUS MATERIALS

Vegetable gums and viscous materials are utilized in foods as stabilizing or improving agents. These materials give viscous solutions, so that the dispersion state of foods can be improved by adding these materials. Although the viscosity of dilute gum acacia solution can roughly be expressed by Einstein's equation, the concentrated solution generally shows visco-elastic properties. It has been shown that the viscosity of gum acacia solution is influenced by the addition of ethyl alcohol, formaldehyde, acetone, and various electrolytes. The pH value also affects the viscosity, with the maximum effect being shown at pH 7. The viscosity of gum acacia solution is lowered by mixing with salts or other gum materials. For example, gum tragacanth decreases the viscosity of gum acacia solution, and then the added gum tragacanth precipitates from the solution. This phenomenon can be utilized for purifying gum tragacanth, although the presence of hydrocarbon, sucrose or starch in the solution protects the gum tragacanth from precipitation.

A study of the electroviscous effect in gum acacia solution shows that the volume fraction of dispersed phase is inversely proportional to the gum concentration, and that with decreasing concentration the gum molecules expand gradually in the medium due to a decrease in the attractive force between the sugar components in the molecules. Although gum acacia is a linear polymer, its behaviour resembles that of guar gum and locust bean gum, which are branched polymers, i.e. gum acacia solution behaves as a branched polymer. This phenomenon is demonstrated by the fact that the viscosity of 20% gum acacia solution is nearly equal to that of 0.5% locust bean gum. Similar behaviour is observed with solutions of alginic acid salt, which is also a linear polymer. In the case of gum tragacanth, the molecule is very extended at pH 8, and the system shows a maximal viscosity as a linear polymer. The sensitivity of gums to pH is influenced by the presence of carboxyl groups or sulfuric acid groups in the gum molecule. The molecule of guar gum is neutral, so that the viscosity of the solution does not depend upon pH, but the viscosity decreases with increasing concentration of strong alkali due to decomposition of the complex molecules. Glucomannan, which is a type of polysaccharide, forms viscous solutions in alkali, and then the viscosity gradually decreases with increasing setting time because glucomannan starts to precipitate in alkali solution.

Gum tragacanth exhibits structural viscosity in solution when its concentration is above 0.5%, i.e. the shear rate of such a solution in a capillary tube is not proportional to the shearing stress. A solution of locust bean gum also shows structural viscosity, but one does not observe non-Newtonian flow with gum acacia solutions even at such high concentrations as 30%. It seems that the molecules of gum acacia assume a spherical configuration in the solution.

Agar can be isolated from seaweeds such as Gelidium, Gracilaria or Ahnfeltia plicata, and it is perhaps a linear polymer of galactopyranose. The gelation of agar solution is protected by the presence of free acid, but divalent ions promote gelation of the solution even at low agar concentration.

There is a strong tendency to gel at pH 8–9, but alcohol or other organic solvents check the gelation of agar solution by promoting dehydration. Although agar gels transform to sols at about 90 °C, the solution starts to gel at about 38 °C when the solution is cooled.

Table 3.9 quotes various data on the vegetable gums and viscous materials. Table 3.10 shows the effect of acids, alkalies, salts and temperature on the viscosity and gelation properties of solutions of vegetable gums and viscous materials.

Alginic acid salts and carrageenin, which are extracted from seaweeds, are often utilized for conferring viscous properties on ice cream, whip cream and soup. In particular, it has been reported that carrageenin is useful as a stabilizing agent for chocolate milk due to interaction between the carrageenin and the milk protein.

TABLE 3.9

Characteristics of edible gums [33]

Popular name	Raw material	Chemical remarks	Main residue	Viscosity[a]
Agar-agar	Seaweed	Mixture of poly-saccharides	D-galactose, sulfate 3,6-anhydro-L-galactose	Gel
Algin	Brown algae	Polyuronic acid	D-mannuronic acid, L-glucuronic acid	1,800
Carrageenan	Red algae	Polysaccharide ester sulfate	D-galactose, sulfate 3,6-anhydro-D-galactose	225
Guar gum	Seed of bean family	Polyhexsose	D-mannose, D-galactose	3,000
Gum acacia	Secretion from a tree	Ca, Mg, and K salts of arabic acid	D-galactose, L-arabinose, L-rhamnose, D-glucuronic acid	Low
Gum tragans	Secretion from a shrub	Mixture of complex acid, polysaccharide and neutral araban	L-arabinose, D-xylose, L-fucose, D-galactose	3,200
Karaya gum	Secretion from a tree	Complex acid of polysaccharide	D-galacturonic acid, L-rhamnose, D-galactose	2,300
Locust bean gum	Seed of a tree	Galactomannan	D-galactose, D-mannose	2,750

[a] centi poise in 1% solution at 25°C.

TABLE 3.10

Physico-chemical properties of edible gums [33]

Popular name	pH	Gelation	Effect of reagents			Thermal effect
			HCl	NaOH	Salts	
Agar-agar	7	Yes	Decrease of viscosity	Increase of viscosity up to pH 8.5, then decrease	Little affected	Rigid gel up to 92°C
Carrageenan	7	Yes	Decrease of viscosity	Decrease of viscosity	Prompt gelation	Sol⇌Gel at 38°C
Guar gum	7	No	Little affected	Little affected	Gelation	Decrease of viscosity
Gum acacia	5	No	Decrease of viscosity	Increase of viscosity up to pH 7	Gelation	Decrease of viscosity
Karaya gum	4.6	No	Decrease of viscosity	Increase of viscosity	Decrease of viscosity	Decrease of viscosity
Gum tragacanth	5.5	Yes	Decrease of viscosity	Increase of viscosity up to PH 8, then decrease	Little affected	Decrease of viscosity
Locust bean gum	5.3	No	Increase of viscosity	Decrease of viscosit; at low concentra-tion Increase of viscosity at high concentration	Gelation	Increase of viscosity up to 70°C

3.2.4. GELATIN

Gelatin is a polymer of amino acids, and it is derived from collagen by hydrolysis with alkali or acid. Collagen, which is found in a fibrous state in animal skin or bone, is a major structural protein of connective tissue. The molecular weight of gelatin ranges from 10000 to 60000, but it is not possible to prepare a gel from gelatin whose molecular weight is 10000–15000.

Gelation is obtained with gelatin even in systems diluted below 0.5% of concentration. The gel transforms to a sol above the melting temperature, i.e. sol-gel inversion appears in a range of temperature, as follows;

$$\text{Gel} \underset{30\,^\circ\text{C}}{\overset{40\,^\circ\text{C}}{\rightleftarrows}} \text{Sol}.$$

Gelation of gelatin solution may be brought about by the formation of a network structure between the gelatin molecules in the medium because of the secondary binding forces of the molecules. The linkages in the network structure disappear with increasing temperature, so that the van der Waals' forces may play a large part in the linkages. Ferry [34] has suggested that the linkages in a gelatin molecule may arise from five or six sources, which corresponds to the types of amino acid residue.

Diluted solutions of gelatin behave as Newtonian fluids above the melting temperature. The viscosity of gelatin solution gradually decreases in reverse proportion to the heating time. This effect is pronounced at high temperature, or with concentrated acid or alkali, because these conditions promote hydrolysis within the

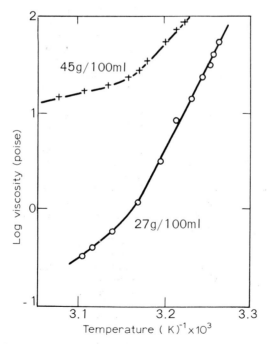

Fig. 3.22.   Dependence of viscosity of gelatinized starch upon temperature.

gelatin molecules. On the other hand, the viscosity of gelatin solution increases in logarithmic proportion to the concentration of gelatin. As shown in Figure 3.22, the logarithm of viscosity can be plotted linearly against the reciprocal of absolute temperature when it is above the melting temperature.

It is possible to obtain Hookean elasticity with gelatin in the gel state using various techniques, such as – (1) bending or extension of a sample (Leick [36]: 1904), (2) compression (Hatschek and Jane [37]: 1932), (3) torsion of a rod like sample (Sheppard and Sweet [38]: 1920), (4) coaxial type rheometry (Kinkel and Sauer [39]: 1925, Cumper and Alexander [40]: 1952), (5) meniscus change of the sample due to compression by air pressure (Saunders and Ward [41]: 1953), (6) Transmission velocity of vibrations in the sample (Ferry [42]: 1941).

The rigidity of gelatin gels increases approximately according to the square of the gelatin concentration. This tendency is, however, much influenced by various factors such as the kind of gelatin, degree of decomposition of the gelatin, thermal history of the system, pH and ionic strength in the medium, and so on.

Ferry [34] has found that a linear relationship can be obtained between the quotient of the rigidity of the gelatin gel to the square of the concentration (G/C^2) and the temperature excepting near $0\,°C$, as shown in Figure 3.23.

Increase in the rigidity of gelatin gel during setting is initially rapid and then it

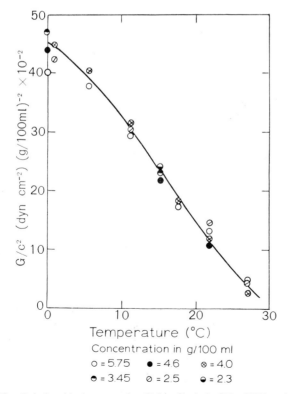

Fig. 3.23. Relationship between the G/C^2 of gelatin ($M=4500$) and the temperature [35].

slows down, so that the rigidity gradually approaches an equilibrium. Gelatin gels generally exhibit a linear relationship between the stress and the strain for short loading times such as a few seconds. When the stress is applied to the sample for a long period of time, the sample shows creep behaviour, i.e. a permanent deformation develops in the sample. This phenomenon may correspond to the dispersion state of the gelatin molecules in the medium such as local crystallization of the orientated molecules, destruction and regrowth of the network structure under the loading, etc.

Major uses of gelatin are in the preparation of deserts, salad paste, and cakes, and especially of marshmallow.

3.2.5. EGG WHITE

Solid components occupy about 12% in egg white, and about 85% of the solid component is protein. Egg white irreversibly transforms to the gel state when heated. One can obtain very strong foamability with egg white by mixing it with sugar or wheat flour. Egg white consists of both thin and thick fluid components; the former is located near the inside of the egg shell and near the egg yolk, and the latter is held between both layers of the thin component. Elasticity does not appear in the thin fluid, but fibrous elastic material can be identified microscopically in the thick component. This keratin-type fibrous protein is insoluble in water, and is generally called ovomucin. It is widely distributed in natural materials.

Egg white starts to coagulate at 55 °C–57 °C, and it is denatured instantaneously by heating above 60 °C, although the coagulation temperature is raised when salts are added or at higher concentration of protein. A high acidity delays coagulation. Sucrose is also one of the components in egg white.

3.2.6. MILK GEL (Consistency of Rennet Milk in the Initial Stage of Coagulation)

Cheese making is started by introducing rennet into acidic milk. When a suitable stiffness has been given to the milk gel, as judged sensorily by an expert, the gel is cut with a curd knife. Scott Blair et al. [43] have attempted to give objective meanings to the judgement of experts, and they have also studied rennet activity and how it influences the gelation of milk. A major component of rennet is the enzyme rennin. Although details of the enzymic action of rennin are still not known, it appears that k-casein in the casein protein is transformed to macropeptide by rennin, while the electrical charges on the polymolecules are masked by enzymic decomposition of the components. The reduction in the electroviscous effect causes a decrease in the viscosity of the casein system.

The presence of calcium also induces coagulation of casein according to the degree of ionization of the calcium. Hostettler and Ruegger [44] have reported that an abnormal viscosity develops in milk when the casein micelles form chain like polymers in milk during the coagulation process. The casein chain polymer confers elasticity on the system through the formation of a network structure. Scott Blair et al. [45] have measured the viscosity changes in the initial stages of the reaction between rennin and sodium casein using an Ostwald type viscometer, in which the sample had been

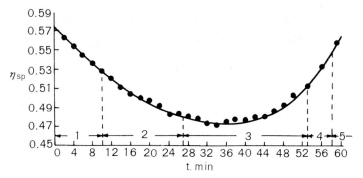

Fig. 3.24.   Specific viscosity change of milk at pH 6.7 and 32 °C.

prepared by adding rennen to a 9% solution of skim milk. The relationship between the setting time and the specific viscosity of the sample at pH 6.7 and 32 °C can be represented by a concave curve, as shown in Figure 3.24. The viscosity decreases first, and then increases with increasing setting time. This process can be sub-divided into five steps, as shown in Figure 3.24, and the first step is a zero order reaction, as follows;

$$-\frac{d\eta_{sp}}{dt} = k_0, \quad \eta_{sp} = k_0 t \tag{9}$$

while the second step can be given by a first order reaction, so that

$$-\frac{d\eta_{sp}}{dt} = k_1 \left(\eta_{sp} - \eta_{sp\min}\right), \quad \ln\left(\frac{\eta_{sp\max} - \eta_{sp\min}}{\eta_{sp} - \eta_{sp\min}}\right) = k_1 t \tag{10}$$

when $t=0$, Equation (10) can be rewritten by $-(d\eta/dt)=k_0$, then

$$-\frac{d\eta}{dt} = k_0 = k_1 \left(\eta_{\max} - \eta_{\min}\right). \tag{11}$$

So it follows that the ratio k_0/k_1 corresponds to $\eta_{\max} - \eta_{\min}$. The above correlation suggests that the rate constant k_1 does not change with the concentration of milk, but that the rate constant k_0 is proportional to the concentration.

The rate constant of the first order reaction k_1 is influenced by the concentration of rennin in the original sample according to the following relation:

$$k_1 = AC_e N \tag{12}$$

where C_e is the concentration of rennin, and A and N are constants, with N depending on the purity of rennin. When the rennin is pure, N is unity.

One can obtain a linear relationship in a plot of the reduced viscosity of the skim milk-water system against concentration, so that the intrinsic viscosity can be measured by extrapolation of the curve to zero concentration of skim milk, as shown in Figure 3.25. Although this intrinsic viscosity decreases as protein is decomposed by enzymic action, the slope of the curve remains constant. Scott Blair *et al.* [45]

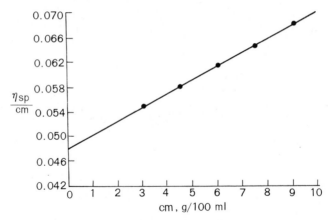

Fig. 3.25.   Intrinsic viscosity of skim milk.

have found a relationship for the rate process of coagulation of rennet milk from the dependence of decreasing viscosity in the initial stage upon the specific viscosity of the original sample,

$$\chi k_1 = 1/\tau \tag{13}$$

where χ is given by

$$\chi = (\eta_{max} - \eta_{min})/\eta_{max}. \tag{14}$$

The rate constant of the first order reaction k_1 depends on the activity of rennet, but it is independent of the quality of casein. The term τ which possesses a dimension of time in Equation (13), depends on both the activity of rennet and the quality of casein.

Ordinary milk, and not skim milk, is used for making cheese, and the expert subjectively judges the time for cutting the curd from the mechanical strength of the surface of a sample. It is necessary, therefore, to evaluate such subjective judgements from the scientific point of view. In order to develop objective measurements for the process of coagulation of rennet milk, Scott Blair has tried to follow the viscosity changes of rennet milk at intervals of 10–15 s using the capillary tube viscometer, which has been mentioned in Chapter 2 of this book.

Figure 3.26 shows the correlation between the height of milk h in the viscometer and the time t. If the plot of $\log h$ against t is linear, the sample is a Newtonian fluid. Ordinary milk generally shows Newtonian flow, but the flow curve of rennet milk deviates from linearity at a time t_c, as shown in Figure 3.26.

Berridge [46] estimated the onset of coagulation in rennet milk from the formation of a white film at the inside of a test tube inclined at 30°, which was filled with the sample and was rotated in a water bath. Figure 3.27 shows the correlation between the reciprocal of rennet concentration and the onset time of coagulation (t_c) of rennet milk, which was determined by subjective judgement of the expert, by the Scott Blair's viscometric method, and by the Berridge's technique, respectively. One finds a reasonably good agreement in the above relationships, although the expert judges

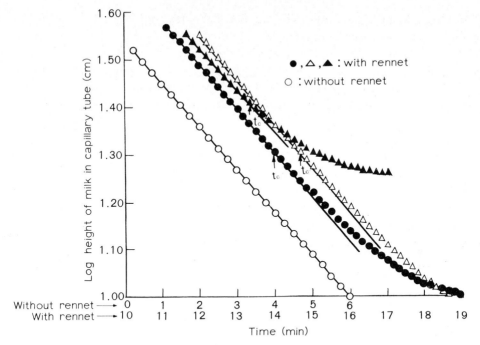

Fig. 3.26.   Flow properties of milk with and without rennet.

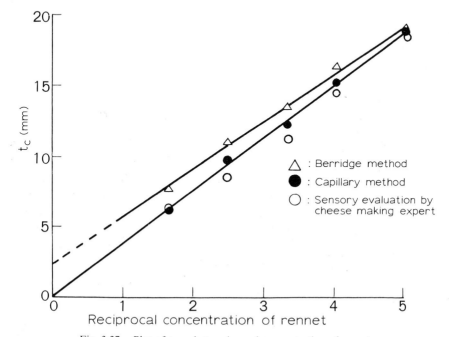

Fig. 3.27.   Plot of t_c against reciprocal concentration of rennet.

t_c at a somewhat earlier time than the Berridge's technique and the viscometric method. The discrepancy between the respective methods may be brought about by the gel structure in the sample breaking down under the large deformation applied in the objective techniques. It is certain that the sample in the cheese vat tends to form a weak gel-structure just after the start of setting, so that the expert judges commencement of the coagulation at an earlier time than do the other methods.

3.2.7. ELASTICITY OF RENNET MILK

Scott Blair and Oosthuizen [47] have tried to obtain information about the dependence of the elasticity of rennet milk upon setting time. The results show that a plot of the elasticity against setting time can be expressed by a sigmoid pattern, and that the final elasticity increases with increasing amount of calcium in the system. If one plots the elasticity G against logarithmic setting time ($\log t$), one obtains a straight line except at the initial stage, as shown in Figure 3.28.

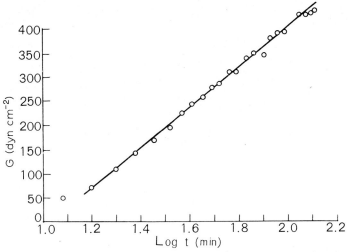

Fig. 3.28. $G \sim \log t$ curve of milk gel in the period of setting.

The term G is only a simple expression of the elastic properties of the sample. It is possible to characterize a number of mechanical parameters for the sample, such as an instantaneous elasticity G_1, retarded elasticity G_2, Newtonian viscosity η_1, and internal friction η_2. These parameters can also be plotted linearly against $\log t$. Intersection of these linear plots on the scale of $\log t$, denoted as I, may correspond to the induction period of gelation, though the term I is only an apparent value because of non-linearity of the plots near the initial stage of gelation. The relationship between the logarithmic induction period $\log I$ and the temperature θ can also be represented by a straight line, as shown in Figure 3.29. An empirical relation for the above linear plots is given by

$$I = e^{B - A\theta}$$

The enzyme rennin decomposes the protective colloid which stabilizes colloidal

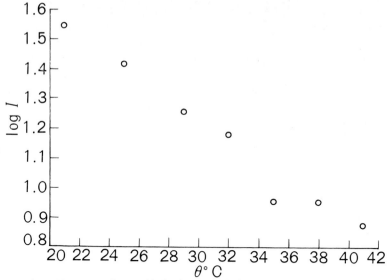

Fig. 3.29.   Dependence of induction period of gelation upon temperature.

casein in milk, so that the para-casein forms a calcium salt and precipitates from the system. The process of coagulation in milk can be separated into two parts; one of them is an enzymic process, and the other is non-enzymic. Berridge has tried to sub-divide the gelation process of milk into two steps, one being a non-gelation process concerned with rennin, and the other a gelation process due to thermal influence, i.e. if milk is heated at various temperatures after treatment with the enzyme rennin at $0°C$, the sample will start to coagulate according to the following rate;

$$R_{(T+t)} = R_T e^{kt} \tag{15}$$

where R_T and $R_{(T+t)}$ are the rates of coagulation at $T°C$ and $(T+t)°C$, respectively, and k equals the temperature coefficient which is given by $R_{(T+t)}/R_T = e^k$. This coef-ficient k corresponds to the factor A which has been derived by Scott Blair et al. as described previously in this section.

In Figure 3.30, the term $G_\infty - G$ is proportional to the amount of casein which is not involved in the formation of gel structure until a setting time t, where G_∞ and G are the rigidities of milk gel at infinite time and time t, respectively. Therefore, the ratio $\log(G_\infty - G)/G$ should change with time. As shown in Figure 3.30, the above relation can be expressed approximately by a straight line for samples which are solutions of milk powder with various concentrations of calcium ions. The plot is approximately defined by

$$dG/dt = kG(G_\infty - G). \tag{16}$$

The plots in Figure 3.30 are, however, not strictly linear but sigmoid, so that the re-lation $G = G_\infty e^{-\tau/t}$ is more applicable, where τ is the period of time for obtaining $G = G_\infty/e$. Figure 3.31 shows the linear relationship between $\ln G$ and $1/t$, in which the slope of the straight line represents the term τ.

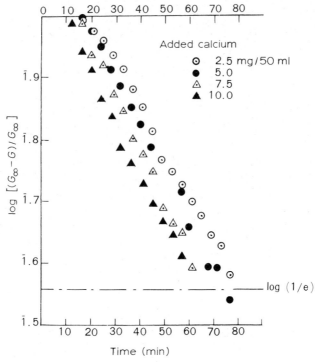

Fig. 3.30.   Gelation curve of powdered milk in aqueous phase.

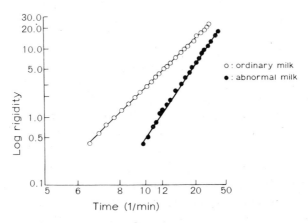

Fig. 3.31.   Log $G - 1/t$ curve of rennet milk.

If one applies and releases the stress periodically in a milk gel, the sample exhibits a hysteresis loop in the plot of stress against strain, as shown in Figure 3.32. The stiffness of the sample becomes progressively lower with repetition of this procedure. Dough also exhibits a similar phenomenon. We denote this phenomenon as the Bauschinger effect, and it is concerned with elastic after-effect.

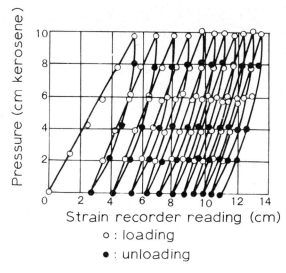

Fig. 3.32. Hysteresis curves for curd.

o : loading

● : unloading

3.2.8. KAMABOKO (Fish Jelly)

Kamaboko is a gel-state food, which is prepared from raw fish by mixing with a small amount of salt, grinding into a mash, and then heating in a fixed size and shape. The gel state is brought about in Kamaboko by heating sol-state mashed fish meat. Kamaboko is one of the traditional foods in Japan which utilizes the physical properties of fish meat proteins, and sensory evaluation has been used to assist manufacture process control and product quality.

(i) *Suwari*

Mashed fish meat gradually starts to coagulate even at room temperature, and we denote this phenomenon as Suwari. Shimizu [50] has studied the stress relaxation of mashed flying-fish meat prepared with 3% salt, and has shown that the tensile force of the sample increases with increasing time at a high temperature, while the relaxation velocity becomes progressively lower. An experimental result on the correlation between Suwari and Ashi* of Kamaboko obtained by Kishimoto and Maekawa [51] who measured the tensile strength of the sample, shows an interesting property of the gelation of Kamaboko, as shown in Figure 3.33, in which the sample has been prepared from mashed Maeso** meat. This mashed meat was treated by heating for fourty minutes at various temperatures ranging from 30 °C to 90 °C, and then a part of the sample was cooled by water (quoted as A in Figure 3.33), while the other part of the sample was heated again for twenty minutes at 90 °C after the

---

* The translator notes that Ashi is one of the sensory terms in Japanese for representing the mechanical properties of Kamaboko, which will be explained in the latter in this section.
** The translator notes that Maeso is one of the bony fishes in Japanese. Up to 50 cm long; found on south Japanese coast, Indian sea, Red sea and Australian coast.

cooling (quoted as B in Figure 3.33). The measurement was made with the above samples 24 h after preparation. In the case of sample A, the gel strength increases with increasing temperature, and it then reaches a constant value above about 50 °C. Sample B, on the other hand, shows a larger gel strength than sample A within a temperature range below 50 °C, but the gel strength reaches a fixed value above 60 °C. Therefore, the critical temperature of gelation in Kamaboko is possibly situated in a range of 50 °C–60 °C.

(ii) *Viscosity of Fish Meat Sols*

It is possible to prepare a fish meat sol from fresh Maeso meat by filtration of the mashed sample, which is prepared by kneading the meat together with one and a half times its own volume of water, aqueous sodium chloride solution is introduced into the mashed sample, and then the sample is allowed to stand for 12 hrs. in the cold. The correlation between the viscosity of the filtered fish meat sol η and the rate of shear D can be represented by $\eta = kD^n$.

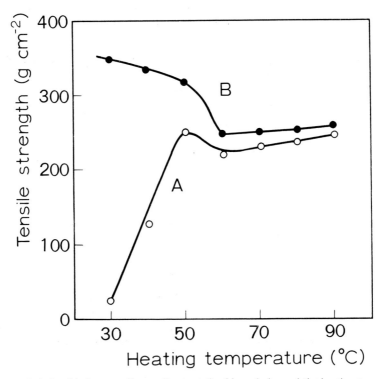

Fig. 3.33.   Relationship between the tensile strength of kamaboko and the heating temperature.

The viscosity of this sol at a fixed rate of shear is a function of the dispersed phase concentration C, such that $\eta = k^m C$. If this sol is heated at various temperatures, and it is then cooled to 5 °C, the viscosity of the sol shows a peculiar pattern according to the measuring temperature, as shown in Figure 3.34. Initially, the viscosity in-

creases rapidly in the temperature range from 30 °C to 40 °C, because the conformation of actomyosin in the system changes before the coagulation. After that
the viscosity suddenly decreases due to aggregation of the coagulated actomyosin.

Gelation in mashed fish meat, which contains a large amount of myosin, may be
due to thermal coagulation of the proteins.

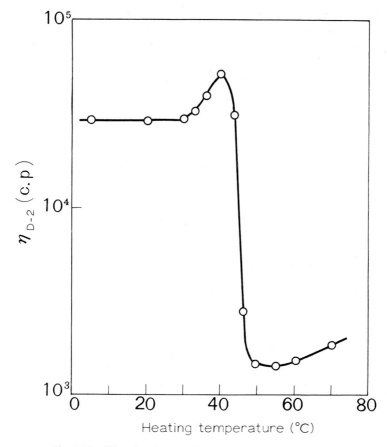

Fig. 3.34. Viscosity change of fish meat sol due to heating.

(iii) Ashi of Kamaboko

The term Ashi has been used by the manufacturers to describe the quality of
Kamaboko, and it is now a popular term for the sensory evaluation of Kamaboko.
It is pointed out by the well trained manufacturer that the term Ashi does not have
a simple meaning, but that it is represents a complex of mechanical sensations in the
mouth such as elastic feeling of Kamaboko on the back teeth, and resistance to
cutting of the sample by the front teeth.

Matsumoto and Arai [53] have tried to obtain information about the correlation
between the sensory term Ashi and its physical meaning using an apparatus, which
was an improved jelly tester. They have shown that the term Ashi can be represented

by the pressure required to compress and fracture the sample with a plunger. Shimizu, H. and Shimizu, W. [54] have shown a consensual correlation for Kamaboko between the sensory evaluation and the gel strength of the sample. This correlation was obtained from measurements of the fracture strength during penetration of a plunger, and also from the relationship between the extended length and the tensile force under a constant velocity of extension with a dumb-bell shape sample.

The correlation data for the stress, strain and Young modulus of seventeen kinds of Kamaboko emphasize that the grade of Ashi does not depend on the Young's modulus of the sample, but that the grade of Ashi improves with increasing stress during the process of fracture, i.e. a high grade of Ashi is closely related to a high extensibility.

(iv) *Correlation between Ashi and Nature of Fish Meats*

The grade of Ashi of reddish fish meat is different to that of the white fish, so that reddish meat is mixed together with white meat in order to obtain a suitable grade of Ashi. The development of Ashi is poor in meat that is not very fresh, because the solubility of myosin protein decreases with decreasing degree of freshness. Although the pH of fresh fish meat is generally around 6, this value increases as the meat putrefies. The increase of pH promotes the decomposition of protein, so that the development of Ashi is retarded.

It is possible to consider the influence of various salts on the development of Ashi in Kamaboko, because Kamaboko is prepared from a protein sol in aqueous sodium chloride solution by thermal coagulation.

Shimizu, H. and Shimizu, W. have studied the dependence of the grade of Ashi upon the concentration of salts, using three kinds of alkali metal salts, and they have shown that there are three critical points of stress against fracture depending on the ionic strength, as shown in Figure 3.35. These critical points may be due to the solubility and salting-out point of protein in the salt solution being affected by the kind of protein in the fish meat tissue. Actually, Kamaboko manufacturers employ salt concentrations of about 0.5 moles/litre of the system for obtaining a reasonable grade of Ashi.

Although factors such as degree of fish freshness, salt concentration, etc. affect the grade of Ashi, it is well known that the important criterion is the amount of soluble myosin. Figure 3.36, which has been taken from Kishimoto's review [55] on the work of Shimizu, shows plots of the logarithm of the soluble myosin protein in mashed meat against the logarithm of tensile force, and there is good linearity for the same kind of fish irrespective of whether it is very fresh or not.

(v) *Viscoelasticity of Kamaboko*

Kishimoto and Hirata [56] have measured the viscoelasticity of Kamaboko by the method of free damped oscillation, and also be creep measurements using Van Holde's apparatus [57]. The relaxation time of Kamaboko τ obtained by the free damped oscillation method at a frequency of 0.1 s^{-1} depends on the manufacturing

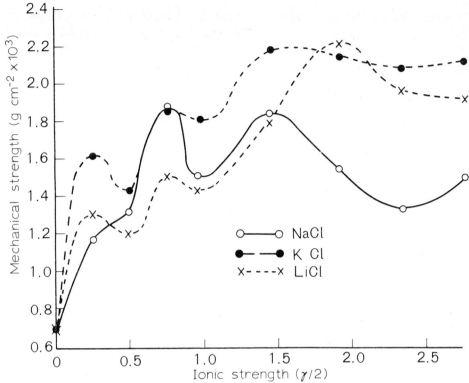

Fig. 3.35.   Mechanical strength of kamaboko vs concentration of metal salts.

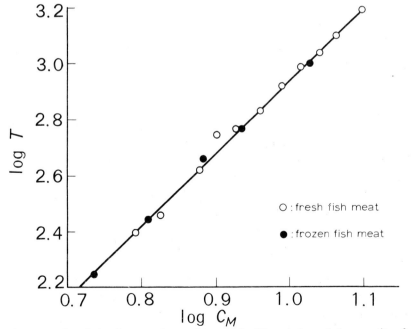

Fig. 3.36.   Correlation between the tensile strength of kamaboko and the quantity of protein in soluble myosin of kamaboko.

source of the sample. This relaxation time can be utilized for estimating the toughness of the sample. Kamaboko is thermorheologically simple, because it is possible to treat the temperature dependence of the viscoelasticity of Kamaboko by the method of reduced variables, as shown in Figure 3.37. One may evaluate the equilibrium compliance J_e as 10^{-5} cm^2/dyn from the result of Figure 3.37, and this value is similar to those for poorly cross-linked rubber like materials.

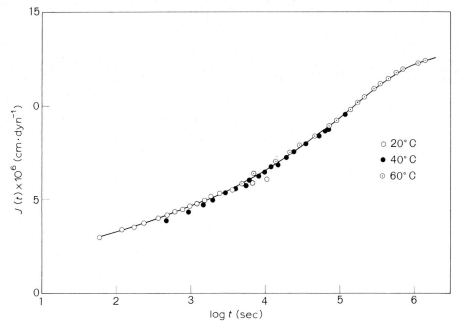

Fig. 3.37.   Composite curve for compliance of kamaboko, standard temperature: 20 °C.

3.3. Fibriform Foods

3.3.1. VEGETABLES AND FRUITS

An important factor affecting the stiffness of vegetables is the cellulosic materials in the tissue. The cellulose materials play a big part in the framework of vegetables. Cellulose is also a major component of plant cells. The cell is covered by a thin film, which is easily separated in the case of mature plants. Young vegetables, however, consists of cells which are covered by a thick wall, and they fit together in a tight block because of the strong intermediate pectin layer.

Lignin contributes stiffness and toughness to the cellulose, and prevents softening and fracture of vegetables during the cooking process. When the amount of lignin in the system is below 2%, the cell swells in the aqueous phase, and polysaccharide molecules dissolve into the aqueous phase from the cell wall. The fibrous state of old beans is brought about by lignification of the epidermal shell. The degree of lignification is influenced by the season, and is promoted by desiccation. Although the amount of fibrous materials in Kohlrabi increases throughout growth, those in carrot or spinach or turnip do not change so much but they do increase pronouncedly during a fixed period of the growth cycle.

Werner *et al.* [58] have tried to obtain information on the properties and weight of fibrous materials in asparagus, and also on the yield of product during the manufacturing process, using a shear-press apparatus, and they have predicted the amount of fibrous materials in the sample from data obtained with the shear-press. Figure 3.38 shows a plot of shear-press value against percentage of fibrous materials in the stalk of raw asparagus. The linear relationship in Figure 3.38 has been obtained by Kramer [61]. As shown in Table 3.11, the shear press value of asparagus is much influenced by the region of the stalk from which it is taken. The value increases as we move from the top of the stalk to the root, e.g. the shear press value at 6.75 in.

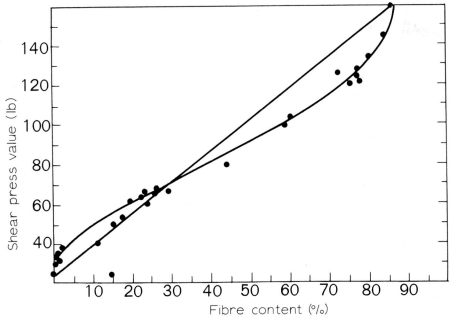

Fig. 3.38. Plot of shear press value of raw asparagus against fibre content in the sample.

TABLE 3.11

Shear press value of asparagus

Distance from the top of stalk (inch)	Number of test	Mean value of shear press (lb)
2.25	14	29.3± 3.5
3.75	14	33.1± ɔ.6
5.25	4	47.5± 7.5
6.75	7	55.5± 7.3
below 1.5 in. from change place	4	108.8±20.0

from the top of the stalk is double that at 2.25 in. below the top. If the sample shows an average shear press value above 60 pounds, it can be predicted that the amount of fibrous materials is so large that the sample is not suitable for use. The shear press value of raw asparagus is intimately related to that of tinned asparagus.

Peas are widely used as a vegetable in Europe and the United States in the form of processed products such as frozen peas, dry peas, or canned peas. Therefore, it is necessary to estimate the maturity which will give the most desireable processed products.

The estimation of maturity grade has been made on vegetables for a long time, and the method consists of physical, mechanical, and chemical techniques, respectively. A tenderometer or maturometer is used to measure the maturity grade of raw peas; these instruments represent mechanical tests. In the case of the processed peas, densitometry can be employed as a physical method, and measurement of the amount of alcohol-insoluble materials is useful as a chemical method.

Lee *et al.* [59] have tried to obtain a correlation between tenderometer results and physical or chemical tests, and they found a good linearity in the plot of tenderometer values against the amount of alcohol-insoluble solid material or starch, and against the density. Densitometry of peas was studied after boiling samples of ordinary peas and also samples which were prepared by using a borer, as shown in Figure 3.39.

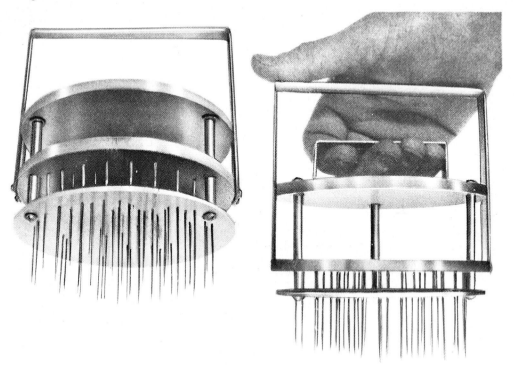

Fig. 3.39.   Borer for testing green peas.

Decker *et al.* [60] have applied an automatic shear press apparatus to vegetables for obtaining information on the maturity grade and the amount of fibrous material, and they obtained results, as shown in Figure 3.40, in which one can find a dependence of recorder response upon the type of vegetable.

Kramer [61] has suggested that it is possible to predict a reasonable harvest time for vegetables using the shear press, because there is a meaningful correlation between the amount of fibrous materials and the stress obtained using the shear press. The U.S. Food and Drug Administration prescribes the amount of fibrous materials in green pea and wax-bean as less than 0.15% from the viewpoint of digestion.

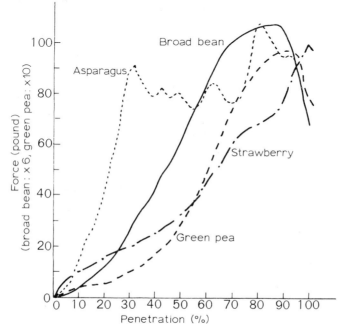

Fig. 3.40.   Results on the shear press test for various vegetables.

The shear press technique for measuring the amount of fibrous material is intended as a replacement for the chemical method in determining the harvest time of beans because of the good agreement between the results of both methods. Table 3.12 shows the theoretical value of the amount of fibrous materials obtained by the shear press technique against the experimental values for various samples, and one finds good agreement between the two sets of data.

Binder and Rockland [62] have measured the change in hardness of lima bean during boiling using a Lee-Kramer shear press (KSP). The force required for shearing each part of the bean decreases with increasing time of boiling, as shown in Figure 3.41, where F_c and F_s are the forces required for shearing an embryo bud and a bean pod, respectively, and W_c is the work required for shearing the whole part of bean. The values of F_s indicates increasing softness as the boiling time increases, although

TABLE 3.12

Relationship between the stress of shear press and the fibre
content in the sample

Size of the sample	Stress (pound)	Fibre content (%)	
		Predicted	Observed
1~3	1.375	0.22	0.20
	1.150	0.07	0.02
4	1.235	0.13	0.11
5	1.455	0.28	0.24

the values of F_c and W_c do not change so much after 20 min of boiling. These
softening rates are proportional to the weight of the sample.

   Fruits are generally preserved by a process such as tinning or salting, and an
important point is to select samples with suitable grades of maturity in order to avoid
the problem of hardness in the products. Deshpande and Salunke [63] have measured
the hardness of apricot and peach so as to obtain information on the maturity and
change of hardness during storage. At the same time they determined the chemical
compositions of the samples e.g. water-soluble solid materials, acid concentration,
the ratio of the amount of water-soluble solid material to acid, sugar concentration,
the ratio of sugar to acid, total amount of pectin material, volatile materials, etc.

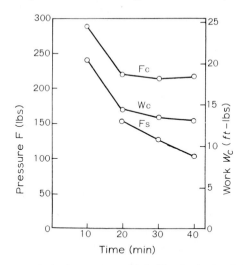

Fig. 3.41.   Softening process with boiling for haricot bean.

Hardness, acidity and the amount of pectin material in the sample, which was
harvested at a suitable level of maturity, decrease during the process of fermentation
at 70 °F, while the amount of volatile materials increases. Deshpande and Salunke
have also suggested that it is possible to prepare a good quality fruit product by
fermentation at 70 °F using raw material whose optimum harvest time can be

determined by applying pressure of 10–15 pounds or 12–13 pounds in a compression type testing machine.

Small fruits such as cherry etc. show hardness changes during storage in salt, and a thrust type testing machine [64] is used for measuring the hardness of such fruits, as shown in Figure 3.42, in which part A possesses an indicator for observing the scale and part C is the moving bar suspended by a spring. The lower end of the bar is connected with a tip (0.08 in. diameter). The hardness of the fruit can be measured by releasing the tip into the sample. Figure 3.43 shows plots of the hardness of salted cherries against storage time; symbol 1 in the figure corresponds to the hard

Fig. 3.42.   Thrust type testing machine [64].

sample, and symbol 2 is for cherries which contain 0.1% pectinase, from which it follows that pectinase plays a big part in the softening process of salted cherry.

Shallenberger and Moyer [65] have measured the hardness of thin slices of tinned apple using the Magness-Taylor apparatus, and they have reported that the hardness is affected by the maturity grade of the original apples, and also by the steaming temperature during processing. The above results indicate that when apples are harvested late, the apple becomes increasingly soft, while the tinned slices of such apple show undue hardness. Factors affecting the hardness of tinned apple slices are the ratio of the amount of alcohol-insoluble solid components (AIS) to the total amount of solid components (TS), and also the heating temperature; the hardness increases with increasing ratio of AIS/TS, and with increasing process temperature. In the case of a low ratio of AIS/TS, the hardness of tinned apple slices increases initially and then it decreases at a high processing temperature. The hardness increases exponentially with increasing processing temperature when the ratio of AIS/TS is high.

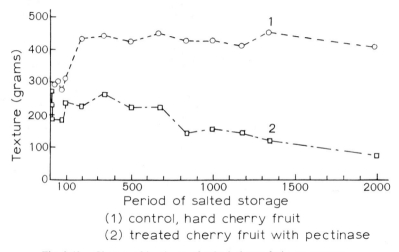

(1) control, hard cherry fruit
(2) treated cherry fruit with pectinase

Fig. 3.43.   Change of hardness of salted cherry fruits on storage.

Hills *et al.* [66] have studied the yield and quality of tinned cherries prepared from bruised material, and they have pointed out that the yield and quality of the tinned product are not affected by the degree of bruising, but they are much influenced by the processing procedure following harvesting. If the cherries are tinned immediately after the harvest, the hardness of the product as measured by tenderometer is low, but the yield of product decreases. They have also emphasized that a suitable hardness and good yield of processed cherries can be obtained by storing the fresh raw sample for 15 h at 18.3°C, but that when fresh cherries are stored for 15 h at a low temperature (1.7°C), the hardness and yield of product are the same as for those prepared directly from the fresh raw material.

Strohmaier [67] has compared the hardness of skin and tissue of frozen apricot with those of fresh and tinned apricot using a penetrometer, and has shown that the skin of frozen samples is not harder than that of fresh apricots due to a softening reaction during storage. His morphological study with a microscope suggests that the cell wall of tinned apricot skin partially swells and dissolves into the bulk under the influence of the thermal treatment.

3.3.2. MEATS AND MEAT PRODUCTS

Meats and meat products are generally expensive foods, so that they are required to be of high quality. The texture of meats and meat products plays a big part in the overall quality.

The relationship between the amount of connective tissue and the relative tenderness of beef has been studied by Mitchell et al. [68], and it has been reported that the tenderness of beef is influenced by the age and breeding conditions of cattle, although they are not absolute factors affecting the tenderness of beef. It is generally known that if a slice of meat is beaten, the slice becomes increasingly tender. Beef-steak is prepared by beating slices of meat, not only because it gives tenderness to the steak, but also because it becomes easier to masticate. Mitchell et al. have pointed out that beating meat affects the amount of collagen and elastin in the meat tissue, although this view is open to doubt.

Muscle consists of many fibrils whose bundles are located in parallel in the network structure of connective tissue. These fibrils consist of elastic proteins, and they show gel-like properties. Fat components are distributed in the muscle capillaries, and the amount and distribution of fat are affected by the location or the breeding condition of the individual animal. The fibrils are not always important to the texture of meat, but the tenderness of meat is much affected by the toughness of the epimysial muscle, by the fat content, by the distribution of fat, by the quality and amount of collagen, and so on.

Tressler et al. [69] have tried to measure the tenderness of meat using two separate instruments; one was an ordinary gauge for determining the compression of tyres, and the other was an improved version of the standard type penetrometer of New York Institute. These instruments gave meaningful data about the tenderness of meat, so that they were useful for comparing the relative tenderness of meat slices with each other. For example, it was observed that the tenderness of quick frozen meat is quite different to that of ordinary meat, that the tenderness increases with increasing period of frozen storage, and that the final tenderness of frozen meat is about 20% higher than that of ordinary meat.

The taste of meat is influenced by the water content of the muscle tissue, and the quality of meat depends upon the taste. Child [70] has designed a compression apparatus for pressing out juice from roast beef, and has established a standard for determining the juiciness of meat. Accordingly, he measured the amount of meat juice and the concentration of dried components, and he also made chemical analyses of the meat juice. The mean ratio of the meat juice to the amount of water in adductor muscle of roasted beef was found to be independent of the period of time for which the sample was pressed, although the average nitrogen content of the juice was affected by the pressing time, and it was much higher in the sample pressed out for five minutes than in that pressed out for twenty minutes. However, these tendencies were not influenced by the point of origin of the meat.

Paul et al. [71] have pointed out that the tenderness of beef on the market does not correlate with the juiciness, water content, and fat content, but that it is intimately related to the fibrous properties of meat. On the other hand, Cover et al. [72] have suggested from sensory evaluation data and shearing tests using a Warner-Bratzler testing machine that one can barely recognize a relationship between the tenderness of beef and the amount of fat in the sample.

Because the results obtained from various test methods for juiciness are not always proportional to those obtained by sensory evaluation Tanner et al. [73] have tried to develop objective methods which would give meaningful relations with sensory tests. Table 3.13 shows the relationship between the juice content obtained by the compression test and the sensory evaluations by a panel. The evaluations used in Table 3.13 were standardized as follows; 5 points – heavy moisture, 4 points – medium moisture, 3 points – light moisture, 2 points – slightly desiccated, and 1 point – well desiccated. Actually, the results in Table 3.13 were recalculated to obtain average values. Although a number of investigators have reported that it is difficult to obtain a meaningful correlation between the tenderness of meat as measured by the Warner-Bratzler shear testing machine and that obtained by sensory evaluation, Kropf and Graf [74] have made an effort to obtain relationships between objective tests, chemical analyses, and sensory evaluations using samples which were prepared from 334 individual cattle.

TABLE 3.13

Juiciness of pork, beef and mutton

Objective evaluation for juice content (%)	Sensory evaluation		
	Pork loin	Beef limb & loin	Mutton limb & loin
25	1.1	1.7	1.8
30	1.5	2.3	2.3
35	2.0	2.9	2.7
40	2.4	3.4	3.1
45	2.9	4.0	3.5
50	3.3	4.6	3.9

With the attainment of full growth in cattle, the muscle tissue of reddish meat becomes increasingly coarse while its colour is getting dark. With increasing round index of cattle the fat content of meat increases, while the grade of marbling in the meat also becomes high. These phenomena relate to the tenderness of meat. Fattiness of meat relates closely to the grade of marbling in the system, stiffness of lean, colour, tenderness, flavour, amount of ether extracts from loin, and so on.

The correlation factor of the sensory test against the tenderness of meat is 0.53

with a level of significance of 5%, but it is not possible to establish meaningful correlations between the sensorily evaluated flavour or juiciness and the tenderness of meat, because the correlation factors are 0.43 and 0.17, respectively, with a level of significance of 5%. The tenderness of lean relates to the external appearance and the inside state (degree of coating by fat and the marbling pattern), and the juiciness of lean correlates to the tenderness with a factor of 0.53. Kropf and Graf have also shown a good correlation between the mechanical shearing force and sensory evaluated tenderness of meat, and they have mentioned that the measurement of shearing force can be used as an objective method for evaluating the tenderness of meat.

TABLE 3.14

Correlation between the subjective, chemical and sensory evaluations for characteristics of beef

1 Length of cattle	2 Round Index	3 Covering of fat (subjective)	4 Marbling (subjective)	5 Tissue of red meat (subjective)	6 Hardness of red meat (subjective)	7 Colour of red meat (subjective)	8 Hardness of bone (subjective)	9 Weight of cattle	10 Sensory favour	11 Sensory softness	12 Sensory flavor	13 Sensory juiciness	14 Mechanical shear value	15 Loin extracted with ether	vs.
	0.05	0.06	0.34	-0.77	0.12	-0.49	-0.76	0.84	-0.55	-0.07	-0.31	-0.03	0.18	0.57	1
		0.89	0.87	0.36	0.90	0.66	-0.19	0.44	0.07	0.53	0.40	0.35	-0.83	0.48	2
			0.84	0.20	0.87	0.48	-0.30	0.35	0.27	0.66	0.48	0.24	-0.79	0.61	3
				0.03	0.90	0.42	-0.49	0.60	-0.03	0.62	0.26	0.37	-0.79	0.74	4
					0.24	0.84	0.64	-0.43	0.41	0.29	0.36	0.29	-0.54	-0.28	5
						0.56	-0.43	0.37	0.07	0.78	0.23	0.34	-0.85	0.65	6
							0.40	-0.08	0.21	0.35	0.45	0.25	-0.74	0.06	7
								-0.50	0.23	-0.32	0.31	0.02	0.06	-0.64	8
									-0.53	0.06	-0.04	0.20	-0.20	0.59	9
										0.53	0.43	0.17	-0.27	-0.07	10
											0.23	0.53	-0.78	0.54	11
												0.18	-0.39	0.26	12
													-0.54	0.24	13
														-0.47	14

In order to measure the tenderness of meat, we have a grinding method which was used by Miyada and Tappel [75]. This method appears as one of the masticating techniques in Sale's classification [76], and it is designed to measure the energy required for grinding meat. Bockian et al. [77] have observed the tenderness of roasted meat using the same method, and they have shown that the experimental results are closely related to sensory test data.

Miyada's grinding technique has been improved by Simone et al. [78] who used an electrical transducer for detecting the resistance encountered during grinding. However, it has been reported that Simone's method is not accurate in comparison with the sensory test, and that the Warner-Bratzler shear testing machine is much better than the grinding method for obtaining a correlation with the sensory tests which has good reproducibility.

Sperring *et al.* [79] have developed a quick technique for accurately measuring the tenderness of both raw and cooked meat with a small amount of sample by improvement of the press juice technique which was designed by Tanner *et al.* [80].

TABLE 3.15

Correlation factor on the stiffness (Longissmus dorsi)

	Group I			Group II
Setting period of heated sample	10 days	10 days	10 days	10 days
Setting period of raw sample	3 days	7 days		10 days
Sample No. →	35	22	57	35
Correlation between the sensory evaluation and Warner-Bratzler shear apparatus (heated sample)	–	–	−0.711	−0.595
Correlation between the sensory evaluation and the compressibility (heated sample)	–	–	−0.355	−0.615
Correlation between Warner-Bratzler shear apparatus and the compressibility (heated sample)	–	–	0.479	0.337
Correlation between Warner-Bratzler shear apparatus (heated sample) and the compressibility (raw sample)	0.785	0.004	–	0.226
Correlation between the sensory evaluation (heated sample) and the compressibility (raw sample)	0.899	−0.017	–	0.199
Correlation between the compressibilities of heated and raw samples	0.415	0.016	–	–

As has been described in the outline of Sperring's method in Section 2.2.4 of Chapter 2, a meat slice of half inch thickness was bored cylindrically by use of a cork type borer, and then the sample was compressed by a plunger in a cylindrical vessel. Tests were usually made at room temperature, and in the case of roasted samples the meat was treated in an oven at 325 °F to obtain a meat temperature of 140 °F. The tenderness of meat was represented by the pressure, which was determined by pressing out meat juice through a 0.3 cm diameter hole at the base of the apparatus, and the tenderness value was obtained as an average of the results from three samples. The cattle samples were divided into two groups such as Group I and II, and the raw samples were used for measuring the tenderness 3 days and 7 days after slaughter. The sample from Group II was provided 10 days after slaughter, and the respective roasted samples from both Groups I and II were also prepared 10 days after slaughter. The roasted samples showed an excellent correlation between the sensory test, the Warner-Bratzler's shear test, and the compression test respectively, but the compression test for the raw samples of 7 days old and of 10 days old did not correlate with either the sensory or Warner-Bratzler tests. The tenderness of 3 days old raw samples as measured by the compression test corresponded to that of 10 days old raw samples as measured by both sensory and Warner-Bratzler tests, and it was also deeply related to that of 10 days old roasted sample as measured by the compression method. Therefore, there appears to be a relationship between the experimental values of the compression test with 3 days old raw beef and the various evaluations of the tenderness of 10 days old samples.

Tuomy and Lechnir [81] studied the dependence of tenderness, shear properties and shear compression value of beef upon cooking temperature and time. Disk-shaped samples were subjected to quick rates of heating and cooling, in which the employed heating temperatures were 140, 160, 180, 190, 200, and 210°F, respectively, and the testing times fell within seven hours. The stiffness of beef increased, as the cooking temperature increased, so that the temperature was clearly involved in stiffening the meat. When the meat was stored at a temperature below 180°F, the stiffness of the sample was not dependent on the storage period, but it was influenced by the storage temperature. However, the stiffness of meat became progressively higher with increasing storage time above a temperature of 180°F. Table 3.16 shows the results of quality tests on meat as obtained by four independent methods, and the values quoted in the table represent the average of five independent measurements.

Meat is possibly initially stiffened by heating because of thermal denaturation of protein. One may find, however, a prolongation of the tenderness of meat even when it is subjected to a relatively high temperature. Some components in meat, with the exception of collagen, may contribute to prolongation of the tenderness. If one heats raw meat at a temperature of 65–72°C just before the desiccation process, the protein in the desiccated sample may be more or less in the non-coagulated state, and the addition of hot water may bring a little stiffness to the system. Also, an initial heating of the meat at a temperature of 65–72°C does not soften the connective tissue of meat, so that the hardening effect due to the components of connective tissue still remains in the restored sample. Such phenomena are important in relation to frozen desiccation of the system.

Ritchey et al. [82] has studied the tenderness of connective tissue in cattle muscle by measuring the collagen content and by sensory tests and he has clarified that long-issmus dorsi contains a smaller amount of collagen-form nitrogen than biceps femoris muscle, although the ratio of transformation from collagen to gelatin is the same in both the above muscles. He has found, however, that the significant difference between cattle appears in the ratio of transformation to gelatin from both collagen-form nitrogen and collagen in the cooking procedure. It has been shown by sensory tests that both muscle of longissmus dorsi and biceps femoris become increasingly tender with increasing processing temperature.

Ritchey has also investigated juiciness, components which play a part in the tenderness of meat, shearing force, elongation properties of fibrous muscle, amount of collagen, etc. with four types of beef such as longissmus dorsi (LD), biceps femoris (BF), semimembranosus (SH) and semitendinosus (ST), respectively. All the samples were cooked as steaks with a thickness of 2.5 cm and at a temparature of 61°C or

80 °C. It was found that the tenderness of steak prepared from LD showed a different temperature dependence than the other samples, because the fibrous components in LD are in close contact with each other so that tearing the sample is difficult, and that the meat of ST becomes hard on heating at both temperatures of 61 °C and 80 °C. Also, that LD contains a small amount of collagen, but that BF consists of thick and coarse tendons.

TABLE 3.16

Comparison of quality evaluations

Temperature (°C)	Heating period of time (hours)							
	0	1	2	3	4	5	6	7
Softness[a]								
60	5.9	5.7	5.8	6.0	6.0	5.6	5.9	5.9
72	5.1	4.8	4.7	4.4	4.2	4.5	4.9	4.7
83	4.5	4.1	5.2	5.0	5.7	5.3	5.7	6.1
87	4.4	4.5	4.8	5.5	5.7	6.4	6.3	6.5
94	4.2	4.8	4.9	6.3	6.5	6.8	6.8	6.5
100	4.0	5.9	6.9	7.4	7.6	7.7	7.9	–
Shearness[a]								
60	5.7	5.4	5.5	5.6	5.5	5.2	5.7	5.4
72	5.4	5.1	4.9	4.6	4.6	4.6	5.0	5.1
83	4.3	4.4	4.9	5.3	5.7	5.8	5.8	6.5
87	4.4	4.7	5.9	5.9	6.6	7.1	7.0	7.2
94	3.9	4.7	5.9	6.5	7.0	7.4	7.7	8.0
100	3.9	6.1	7.2	7.8	8.1	8.3	8.4	–
Shear compressibility[b]								
60	328	349	378	394	371	398	368	345
72	382	429	422	427	411	406	391	343
83	456	440	346	316	304	287	321	265
87	402	374	296	250	218	191	176	157
94	460	335	251	249	167	126	130	123
100	491	276	219	154	151	167	136	–
Flavour[a]								
60	5.9	6.0	5.6	5.7	5.6	5.8	5.8	5.8
72	5.8	5.4	5.6	5.5	5.2	5.1	4.9	5.0
83	5.9	5.4	5.8	5.5	5.6	5.8	5.6	5.5
87	5.4	5.1	5.3	5.4	5.2	5.2	4.9	5.1
94	5.5	5.7	5.6	5.8	5.5	5.7	5.4	5.5
100	5.4	5.8	5.8	5.8	5.5	5.8	5.6	–

[a] Sensory test with nine degrees of scale.
[b] Mean shear compressibility of raw meat is 122 pounds.

A number of studies have been made on the process of rigor mortis. For example, measurements have been made on the extensibility and elasticity of smooth muscle fibres of various sizes. Recently, Briskey et al. [83] designed an apparatus, the so-called rigorometer, which consists of artificial teeth made from lucite, and it is able to apply various loads to the fixed muscle fibre at a constant temperature and humidity. They measured the tenderness of smooth muscle fibre (1 cm^2 × 8 cm), which was prepared from white meat of pig immediately after the slaughter, with this rigorometer. The apparatus recorded automatically the process of rigor mortis at intervals of two minutes, as shown in Figure 3.44. The curve in Figure 3.44 can be divided into a phase of diminishing elasticity, and a region from which elasticity is completely absent and the latter region may correspond to rigor. The time dependence of rigor mortis is not influenced by the loading interval or by the size of the muscle sample, but it is affected by the temparature, i.e. the process of rigor mortis becomes slower with reducing temperature. The white fibre of semitendinosus shows a delayed phase for rigor, which is longer than that for the reddish fibre.

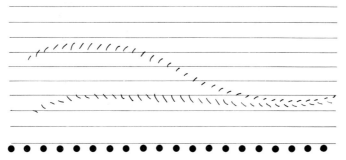

Fig. 3.44.   Recorded chart of rigor meter.

It is possible to deduce from the above results that one of the factors affecting the tenderness of animal muscle arises from the different degrees of contraction of myofibril in muscle. Locker [84] has reported that the degree of contraction may participate in the tenderness of meat. Different lengths of sarcomere in muscle cause strain to the muscle when the carcass is suspended perpendicularly, so that the long sarcomere plays a big role in the softening process of muscle. Herring *et al.* [85] have shown the dependence of extension or contraction of muscle upon sarcomere length using a tensile testing machine. Muscles of both semitendinosus and psoasmajor are deeply involved in rigor mortis, and tensile tests made with both muscles during the process of rigor mortis showed that sarcomere length affected both the induced extension or the contraction during rigor mortis.

Although a number of fresh fish are cooked immediately when preparing dishes for a meal, some of them are preserved in cold storage, in salt, in desiccation, or by other methods, in order to provide materials which can be cooked at any time. Such stored fish are expected to have either a fresh or special taste. It is, however, generally difficult to restore the freshness in stored process fish. This problem will be discussed initially with reference to cold storage.

To measure the hardness of muscle in cold stored or refrigerated fish by the apparatus used for ordinary meat is difficult, because the length of fibre in fish meat

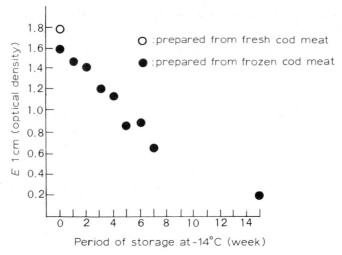

Fig. 3.45.   Plot of optical density of cod muscle juice against storage period of time.

is shorter than that in ordinary meat, and also because the hard connective tissue disperses in fish muscle tissue.

Love [86] has attempted to compare the hardness of refrigerated cod meat with the results of optical density measurements on the cod meat emulsion, the emulsion being prepared by immersion and homogenization of the cod meat in formaldehyde. Figure 3.45 shows the results obtained using refrigerated cod at − 14 °C. The optical density of the emulsion decreases with increasing storage time, and the amount of water soluble protein decreases with homogenization.

An inferior quality of refrigerated fish appears to be produced by the fracture of muscle cells due to the expansion of ice crystals in the cell, but Reay and Kuchel [87] have shown that the poor taste of refrigerated fish is caused by a decrease in the amount of soluble protein. It is, however, not clear what role soluble protein plays in the hardness of fish meat.

Hotani [88] has investigated the relationship between the internal friction and the amount of ammonia-form nitrogen in refrigerated meat of tunny, using an oscillation technique for measuring the internal friction. The experiments were made with three kinds of tunny such as Yellow fin, Sword fish and Albacore, and the samples were preserved at a fixed temperature (− 17 °C) after refrigeration at various cooling rates, but the internal friction was not affected by the rates of refrigeration nor by the storage temperature.

Love and Elerian [89] have concluded that the denaturation of protein in cod meat caused by refrigeration is accelerated by an increase of salt concentration, which freezes water in the protein. Therefore, the denaturation of protein can be checked by immersing the sample in glycerine immediately before refrigeration so as to prevent the crystallization of water in the protein.

Olley and Duncan [90] have analysed the increased amount of free fatty acid in fish meat resulting from refrigeration using a gas-liquid chromatograph. Although it was difficult to obtain reliable information on the nature of the free fatty acid, on the chemical composition of phospholipids, and on the neutral fat in various refrigerated samples, they have recognized a tendency for the rate of protein denaturation to increase with increasing amount of free fatty acid in the system.

Fresh fish meat generally shows a toughness and tightness, while the meat of old stored fish is soft. This tenderness in the stored meat may be brought about by microbiological digestion or enzymatic autolysis during storage at a relatively high temperature, but the decrease of hardness in the meat is actually caused by such complicated factors that it is difficult to find any theory which will explain this phenomenon.

Love and Elerian [89] have measured the toughness of refrigerated cod meat. They have pointed out that myofibril is broken by homogenization of fresh cod muscle, and that the amount of fractured myofibril is lowered by denaturation of protein in the refrigerated muscle. The hardness of myofibril increases, while its solubility decreases due to the denaturation of protein. Love *et al.* have also illustrated that the reciprocal of the optical density $(1/r)$ of cod meat emulsion correlates with storage time t, as follows;

$$\ln \frac{1}{r} = kt, \tag{17}$$

where k is the constant, so that reciprocal of the optical density can be adopted as a toughness index. Dried fish meat is generally harder than refrigerated fish meat, and the degree of hydration in restored dried fish meat is very low. The relationship between the strain and the stress of dried cod meat is shown in Figure 3.46, in which the strain is represented by the ratio of stretched length to the original length. Dried samples require to be subjected to strains which are three times larger than for ordinary meat samples.

It has been shown [91] that 'Surume', which is dried cuttlefish found in Japan, shows rubber elasticity when restored with water. Protein fibre in cuttlefish consist of giant macromolecules, and one may find crosslinkages in the fibres. Such linkages

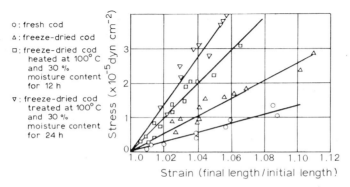

Fig. 3.46.   Stress-strain diagram of cod muscle fibre bundles.

may be formed by the desiccation of cuttlefish, and the molecular weight of the segment between the linkages is about 6000–7000. Figure 3.47 shows a stress relaxation curve of restored 'Surume'. The stress of 'Surume' relaxes very quickly when the fish is restored with aqueous urea (2.5 M) solution, because of the dissolution of hydrogen bond, and stress relaxation can be promoted still more in the presence of 0.1 M hydrochloric acid due to fracture of the cross-linkages in the system, but the effect of $NaHSO_4$ is negligible small. Therefore, the disulfide bond between the protein molecules does not play a big part in the elastic properties of 'Surume'.

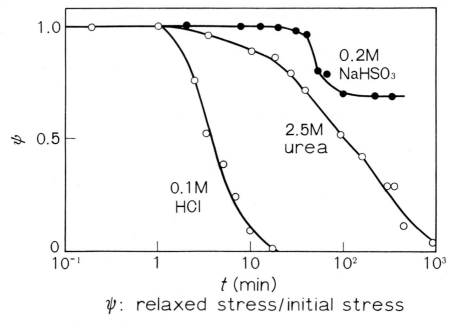

Fig. 3.47.   Effect of additives on the stress relaxation of restored surume (dried cuttlefish).

3.4. Cellularform Foods

3.4.1. POTATO

The quality of potato is usually assessed by the condition of the tissue in cooked potato, i.e. one may require that boiled, mashed or baked potato possesses a mealy tissue which can be identified as brittle and dry when cooked potato is pressed down by fork or spoon.

It is well known that the pliable tissue of potato is surrounded by epithelial tissue, and that it is compactly filled with a number of cells in which starch granules are suspended in the cytoplasm. These cells in the pliable tissue are tightly bound together by connective fibre. The cells and starch granules become increasingly large with increasing distance of the pliable tissue from the epithelial layer.

The cell wall consists of cellulose microfibres, and the intercellular phase is filled with polyuronide which is involved in connecting the cells together. Polyuronide is a kind of pectin material, and it is associated with the cell by means of hydrogen bonds which are, however, broken by heating due to dissolution of the polyuronide. Starch granules swell and dissolve in the cytoplasm by heating, so that the system is transformed into a gel-like state. Softening of the pliable tissue is brought by thermal changes in the intercellular phase, and the connective strength between the cells decreases. The tensile strength of the pliable tissue in native potato is 7 kg cm^{-2}, but in boiled potato it is only 0.3 kg cm^{-2}.

If the binding ability of the pectin material in the intercellular phase is poor, and if the cell contains many starch granules and is surrounded by a strong cell wall, the cell becomes spherical without fracture as result of the swelling pressure at the gelatinization temperature, and the respective cells break away from each other. Such phenomena are responsible for mealy tissue in cooked potato. On the other hand, if the binding ability of the pectin material is pronounced, and if the amount of starch in the cells is small, it is difficult for the cells to disentegrate in cooked potato, since the pliable tissue appears to be sticky or waxy. The mealy tissue is morphologically compared with sticky tissue in Figures 3.48 and 3.49. The stickiness of the tissue is also brought about by the starch gel overflowing from the swollen cell due to fracture of the cell walls, so that it is necessary for the mealy boiled potato to have strong cell walls if this is to be prevented. Reeve and Notter [93] have illustrated from their morphological study that the tissue of boiled potato is affected by the degree of thermal treatment and by the amount of water, and that the water absorbability of dried potato slice prepared from boiled potato is influenced by pretreatments such as heating, freezing, desiccation, etc. Water absorbability of the tissue closely relates to the porosity of potato.

Dried mash potato can easily be restored with hot water, but it is necessary to

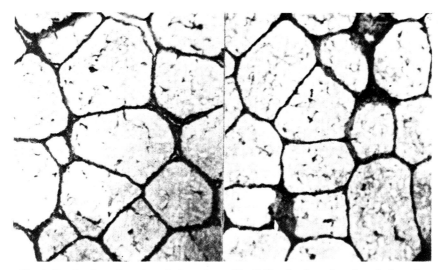

Fig. 3.48. Section of mealy soft tissue in   Fig. 3.49. Section of mealy soft tissue in
'dry' potato (boiling for 60 min) [92].   sticky potato (boiling for 60 min) [92].

provide smoothness and less cohesiveness in order to satisfy the taste. Therefore, raw potato should be preserved in order to obtain low cohesiveness in the tissue, for which the necessary factors are the small amount of amylopectin, large amount of amylose and strong cell walls for preventing fracture of the cell.

Unrau and Nylund [94] have reported that there is a relationship between the relative viscosity of mash potato suspension at a relatively high temperature and the mealiness of mash potato as evaluated by sensory assessment. The relative viscosity also agreed with the amount of starch and with the consistency of mash potato as determined by the Brabender amylograph. Kuhn *et al.* [95] have observed that the relative viscosity change in potato paste during heating is affected by the kind of raw potato; a mealy potato shows a much more rapid increase of the relative viscosity to a high value, and it also shows a relatively low gelatinization temperature. The observation that there is a correlation between the density of raw potato, the water absorbability of cooked patato and the mealiness from the view point of sensory assessment is significant.

3.4.2. RICE AND BOILED RICE

Rice has a long history as a staple food in Asia, especially in Japan, and it is still one of the important foodstuffs. A large amount of rice is also cultivated and consumed in Europe. There are very many kinds in rice, and the characteristics of Japanese rice are quite different from those of rices cultivated in Europe and America. European people generally favour hard and less sticky boiled rice, but Japanese people require boiled rice which is soft and sticky. As shown in Figure 3.50, un-polished rice removed from husk consists of embryo bud and endosperm, which are surrounded by the rice bran layer. The rice bran layer consists of rice skin and a powder layer, and the embryo bud and the rice bran layer are removed from rice grain when preparing boiled rice. Therefore, the nature of boiled rice only depends on the nature of the endosperm, which consists of a large number of cells whose size is about $40 \times 50 \mu - 80 \times 105 \mu$, in which starch granules are found. The starch granule is about $2-3 \mu$ diameter, and it is covered by a thin protein film. The swelling of rice and the consistency of gelatinized starch are subtly influenced by various factors such as the fibrous nature of the cell wall, the mechanical strength and water permeability of pectin material, and the amount of protein.

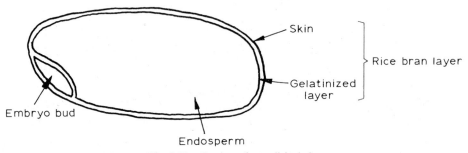

Fig. 3.50.   Section of unpolished rice.

In Japan, rice is usually classified into hard, wine, and soft rices, respectively, according to the morphological examination. The nature of rice can also be predicted from its external appearance e.g. long, medium, and short grains. It has been illustrated by Beachell [96] that the boiled long rice is dry and still possesses the embryo bud, but that well boiled short rice is quite sticky.

The consistency of boiled rice is much influenced by the relationship between the amylose content and the gelatinization temperature. Rao *et al.* [97] have reported that the amylose content is closely related to the swelling of rice, and that the stickiness of boiled rice increases with increasing gelatinization temperature. The short grain shows gelatinization at a relatively low temperature. Deshikacher and Subrahmanyan [98] have compared the swelling of fresh rice grain with that of old grain, and they have shown that the fresh sample appears to give more sticky boiled rice than the old rice. This difference may be brought about by differences in the cell walls between fresh rice and old rice.

Takeo *et al.* [99] have measured the viscoelasticity of gelatinized rice starch using an oscillational rheometer. Both viscosity and elasticity were affected by the starch concentration under fixed conditions of gelatinization, and the viscoelasticity of Japanese rice differed from that of imported rice, as shown in Table 3.17.

Gelatinized starch prepared from imported rice showed a higher elasticity than that prepared from Japanese rice. The viscosity of gelatinized starch prepared from imported rice increases with increasing temperature, but the elasticity decreases slightly, while the viscosity and elasticity of the gelatinized starch prepared from

TABLE 3.17

Viscosity and elasticity of gelatinized rice starch at the temperature of 10 °C

Concentration (%)	Imported rice		Japanese rice	
	Viscosity (poise)	Elasticity (dyn cm^{-2})	Viscosity (poise)	Elasticity (dyn cm^{-2})
5	4×10^2	5.0×10^2	1×10^2	8.5×10^1
10	5×10^3	4.5×10^3	2×10^3	5.5×10^2

TABLE 3.18

Dependence of viscosity and elasticity of gelatinized rice starch upon temperature at 5% concentration

Temperature (°C)	Imported rice		Japanese rice	
	Viscosity (poise)	Elasticity (dyn cm^{-2})	Viscosity (poise)	Elasticity (dyn cm^{-2})
10	4×10^2	5.0×10^2	1×10^2	8.5×10^1
30	4×10^2	3.8×10^2	1×10^2	8.9×10^1
50	18×10^2	3.0×10^2	1×10^2	8.9×10^1

Japanese rice are less temperature dependent. It seems that such phenomena may be related to one of the factors producing lesser stickiness in boiled imported rice.

It has been reported [100] that the viscosity and elasticity of boiled rice are about 10^6 poise and 10^5 dyn cm^{-2}, respectively, as measured by means of the parallel plate plastometer. It is clear that the viscoelasticity of boiled rice correlates with sensorily evaluated stickiness and hardness, and with the overall taste of rice, although the taste of boiled rice also relates closely to various factors such as water absorbability, swollen volume, gelatinization temperature, amylogram pattern, etc.

3.4.3. MACARONI, SPAGHETTI, AND UDON (Japanese Noodle)

Macaroni and spaghetti are prepared from wheat flour dough 'Durum Semolina' by extruding the dough from a thin pipe, and by desiccation of the extruded dough at a mild temperature so as to develop a moisture content equivalent to that of wheat flour.

Japanese noodle 'udon' is also prepared from wheat flour dough. Wheat flour is mixed with aqueous salt solution, and it is then moulded in a thread like shape, and the moulded dough is finally heated in boiled water. Wheat flour consists of starch and crude protein 'gluten', and the elasticity of wheat flour dough increases with increasing amount of gluten protein in the flour, so that it is better to prepare 'udon'

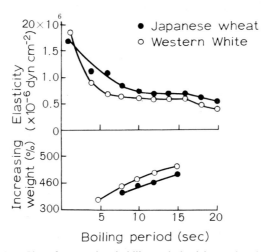

Fig. 3.51. Plot of water absorbability and elasticity against boiling time.

from strong flour, which contains a relatively large quantity of gluten. Shimizu *et al.* [101] have designed a tensile testing machine for obtaining information on the correlation between the elasticity of 'udon' and the boiling period of dough using two kinds of flours prepared from Japanese wheat and Western white wheat. Japanese wheat 'udon' showed little increase in weight during the boiling period because of the poor absorbability of water, and it had a relatively high elasticity, as shown in Figure 3.51. Actually, 'udon' consists of gelatinized starch and thermally denaturated protein, and its mechanical behaviour can be explained from the experimental results on the rate of extrusion under a fixed load by use of a four element mechanical model.

It is also possible to measure the mechanical behaviour of 'udon' using the Brabender extensograph, i.e. the results show that the elasticity and viscosity of boiled 'udon' are 10–100 fold larger than those of dough. The water absorbability of boiled 'udon' also reaches to 400–500% of that of the solid component, and this is pronouncedly larger than for the dough. Although the water absorbability of 'udon' increases with increasing boiling time, the elasticity remains almost constant except at the initial stage of boiling (about 5 min). Therefore, it seems that the network structure due to protein molecules in the system plays a big part in the mechanical properties of 'udon', and that the increase of water absorbability may be brought about by the gelatinization of starch in the system.

Figure 3.52 shows the plots of stress relaxation time of 'udon' against the quantity of crude protein in the samples; all the systems were prepared by boiling for twelve minutes. The apparent value of the stress relaxation time increases as the quantity of protein increases, so that the so-called 'stretch' phenomenon of 'udon', which is caused by heating in boiled water, may be much affected by the protein content.

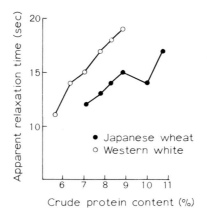

Fig. 3.52.   Relationship between the relaxation time of udon and the
crude protein content in the sample.

Figure 3.53 shows the relationship between the quantity of crude protein and the energy of fracture. With increasing protein content the energy required for the fracture of 'udon' increases. It should be mentioned that the 'udon' prepared from Western white flour is mechanically stronger than that prepared from Japanese flour over a range of low crude protein contents in the system. Harris and Sibbitt [102] have confirmed that the hardness of macaroni becomes increasingly high with protein content. The hardness was also influenced by the climate in which the wheat was grown, and by the kind of grain.

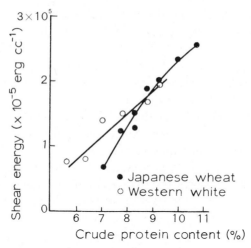

Fig. 3.53.   Shear energy of udon at different crude protein contents.

3.5. Edible Oils and Fats

3.5.1. PLASTIC FATS

Butter, margarine, chocolate, etc., composed of either animal or vegetable oil, are classified as plastic fats. We are accustomed to spreading butter or margarine on bread, and we sometimes prepare cream for cakes with them. Shortening is usually mixed with wheat flour for preparing the raw materials of bread and cake. Chocolate is directly consumed as an article of luxury, or it is used for coating cakes. Thus, the plastic fats are continuously being evaluated for consistency by users and consumers from either the subjective or objective view point. For example, there are two requirements for butter; one of them is ease of spreadability when spreading butter on bread, and the other is the mechanical properties of butter around body temperature. In order to satisfy both requirements, it is necessary to investigate the structure of plastic fats.

(i) *Habit of Plastic Fats*

Plastic fats have a structure in which small fat crystals (a few microns in size) are dispersed homogeneously in the amorphous liquid oil, and a part of the liquid oil is locked in the fat crystals, as shown in Figure 3.54. When a large quantity of fat crystals are dispersed in the system, the crystals form the so-called scaffolding structure, so that the system exhibits a solid state. One may find a yield value in such plastic fats, i.e. the fat never flows when the pressure has a value smaller than the yield value.

However, the system behaves as a fluid, when the pressure exceeds the yield value. If a linear relationship can be seen between the stress and the rate of shear in the flow region of the system, we would classify the system as a Bingham plastic body. As a matter of fact, it is difficult to identify a clear yield value with plastic fats, though one can recognize the upper and lower yield values with a sample. They are also non-linear in plots of stress against rate of shear in the range of both these yield values.

The IOCC (International Office of Cocoa and Chocolate) has attempted to summarize the relationship between stress and rate of shear using a standard chocolate sample, sample pieces being distributed widely to testing laboratories all over the world for obtaining information on the flow behaviour of the sample under fixed conditions using the various types of testing machine. The results obtained are shown in Figure 3.55, in which we find that the rate of shear increases linearly with increasing stress of over 600 dyn cm^{-2}, and that the yield value is about 50 dyn cm^{-2}.

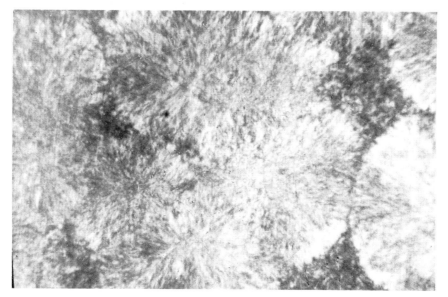

Fig. 3.54.   Microphotograph of plastic fat.

(ii) *Dependence of Consistency upon Temperature*

The melting points of plastic fats are relatively low, and the rheological properties of the systems are pronouncedly affected by temperature over a very narrow range of temperature. The fats consists of tri- and/or di-glyceride molecules, and they show non-Newtonian flow in the melted state. Sterling and Wuhrmann [104] have measured the extrusion viscosity of cacao butter using an Ubbelohde type viscometer, and have shown that the viscosity of cacao butter depends upon the pressure used, as given in Figure 3.56. Although it was not possible to identify the fat crystals in the system with an optical microscope at 30 °C, there were slight indications of a structural viscosity, which may be brought about by either the presence of fine crystals in a sub-micron range or by the asymmetry of glyceride molecules. The structural viscosity of the system became progressively higher with decreasing temperature because of the appearance of fat crystals in the sample.

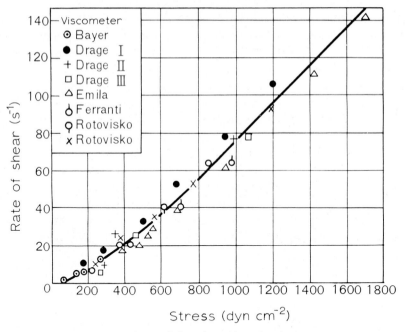

Fig. 3.55.   Stress and rate of shear for molten chocolate (37.8 °C) [103].

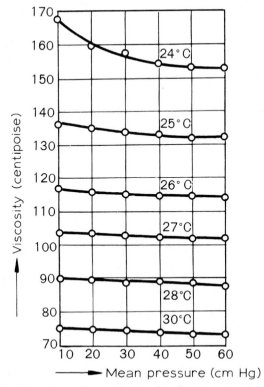

Fig. 3.56.   Dependence of cacao butter viscosity upon extruded pressure.

Micropenetration methods are used for evaluating the hardness of plastic fats. A thin needle is allowed to fall on to the surface of a sample from a fixed height in order to avoid much disturbance to the inner structure of the sample, and then the depth of penetration is measured by means of a dial gauge. It is necessary to keep the temperature of the sample precisely constant as the penetration occurs only in a shallow region near the surface of the sample. Figure 3.57 shows the relationship between the hardness of fats and the temperature as obtained using the micropenetro-meter. Hardened cotton seed oils have a tendency for the temperature dependence of hardness to be greater at low temperature (below 15 °C) than at high temperature (above 30 °C). In the case of butter and lard, however, the temperature dependence of hardness is still pronounced at a temperature of 30 °C, when the first order transi-tion can be seen in the fat crystals, so that the scaffolding structure of the crystals in the system should gradually be weakened. Lard clearly shows a first order transition near a temperature of 25 °C, and the macroscopic structure in the system suddenly transforms at this temperature.

Butter and margarine are required to exhibit spreadability and elasticity when one spreads them on bread. Shortening is usually mixed with flour, or used in the pre-paration of synthetic cream at a temperature of about 20 °C, so that it is desirable to keep the consistency of shortening constant near the prescribed temperature. Davis [105] has tried to obtain information on the viscosity, elasticity and relaxation time of butter from measurements on the permanent deformation and elastic re-covery of cylindrical samples under compression. Butter is prepared from milk, and the composition of milk fat depends on the condition of growth and the kind of cow. Therefore, some degree of fluctuation can be seen in the viscosity and elasticity of

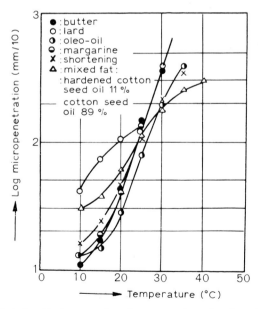

Fig. 3.57.   Relationship between the micropenetration of various fats and the temperature.

the various butters, as shown in Table 3.19. The rheological properties of butter also depend on the method of kneading, so that the crystal form of fat and the macro-scopic structure developed by the fat crystals may be closely related to the rheological properties of butter.

TABLE 3.19

Viscoelasticity of butter

Breed of cow	Working	$\log \eta$	$\log E$	η/E
Short horn	Deficiency	7.08	5.98	1.10
	Normal	7.21	–	–
	Sufficiency	7.28	6.17	1.11
Guernsey	Deficiency	7.51	6.02	1.49
	Normal	7.62	6.12	1.50
	Sufficiency	7.68	6.22	1.46
Australia	–	7.74	6.39	1.35

Hunziker *et al.* [106] have reported from the results of compression tests with cylindrical butter samples that the melting point of butter fat is not influenced by the kind of feed, but that the hardness of the butter increases as the amount of oleic acid in the feed increases. On the other hand, Coulter and Hill [107] have mentioned that the hardness of butter is related to the amount of butter fat and to the iodine-value of the fat in the system, and that butter becomes soft with increasing amount of unsaturated fatty acid.

The spreadability of butter plays an important role in the evaluation of quality, which is subjectively assessed when one spreads butter on bread. Butter is generally consumed at about 16–17°C, at which temperature the correlation between the sensory assessment of consistency and the viscoelasticity of butter has been measured, as shown in Figure 3.58. Normal butter shows an elasticity of about 10^6 dyn cm^{-2}, viscosity of about 3×10^7 poise, and relaxation time η/G of about 30 s.

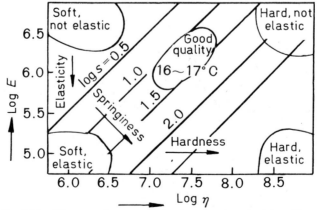

Fig. 3.58.   Diagram of sensory evaluations for viscoelasticity of butter.

(iii) *Consistency and Fat Crystals*

The consistency of edible fat is affected by the form and quantity of fat crystals in the system. Butter, margarine, chocolate, shortening, etc. are generally very soft and fluid immediately after manufacture.

Most edible fats are found in a plastic state at temperatures well below the melting range. The viscosity of the system increases with increasing crystallization of fat, and the non-Newtonian behaviour becomes increasingly pronounced.

Figure 3.59 shows the change in the apparent viscosity of cacao butter against the aging period at 24–25 °C. Microscopic examination of the same material at the same temperature indicates that the fat crystals do not have much influences on the viscosity at relatively high temperature, but the viscosity becomes progressively higher according to the degree of crystallization of the fat in a range of the relatively low temperatures below 25 °C. In the initial stages of aging at low temperature, however,

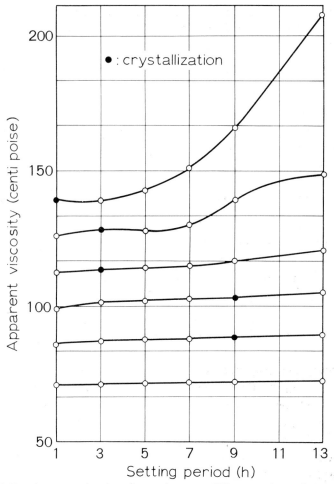

Fig. 3.59.   Apparent viscosity of cacao butter at different setting periods [108].

the viscosity of the system is nearly constant because the crystals are so small that the interaction between them may be negligible.

Feuge and Guice [109] have defined the hardness index (*HI*) using Brinell's hardness meter, as follows;

$$HI = \frac{P(100)}{\frac{\pi D}{2}\left(D - \sqrt{D^2 - d^2}\right)}, \tag{18}$$

where P is the thrust pressure, D is the diameter of the steel ball which is thrust on to the surface of the sample, and d is the thrust diameter on the surface. They have studied the hardness changes during the storage of the sample using the hardness index as a criterion. Figure 3.60 shows the change of hardness index of hydrogenated cotton seed oil, cacao butter and glyceryl tri-stearate against the aging period at about 25 °C. In the case of cacao butter, the sample was solidified by cooling at 15 °C, then it was kept at 21 °C for 2 days, and finally it was stored at 25 °C. The hardness of both the cotton seed oil and the tri-glyceride increases exponentially with increasing aging period, but that of cacao butter increases rapidly after about 20 days aging.

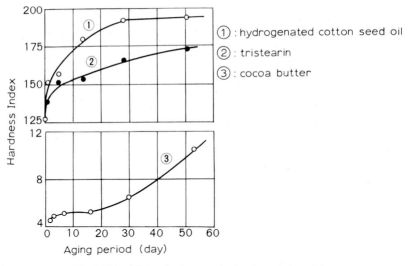

Fig. 3.60.   Effect of aging on the hardness index of fats.

The above results may correlate with those obtained by Sterling and Wuhrmann [104], who reported that a number of fine fat crystals gradually grow from the very many crystal nuclei in cacao butter, but that hydrogenated cotton seed oil and glyceryl tri-stearate easily crystallize into coarse crystals from a few crystal nuclei. On the other hand, some kinds of fats whose melting points are relatively low such as the lauric acid-type confectionary fats, show a high hardness index in an initial stage of storage, and then the index gradually decreases with increasing storage period, as shown in Figure 3.61. This phenomenon may be brought about by the fat component firstly crystallizing as fine crystals, and then transforming slowly to coarse ones due to aggregation of the fine crystals.

Huebner and Thomson [110] have investigated the correlation between the crystal growth of mixed fats in butter and the temperature by studying the relationship between the degree of thermal treatment and the spreadability of butter using a spreadmeter. The butter samples were kept at various temperatures ranging from $-21\,°C$ to $13\,°C$, and they were then evaluated for spreadability at a fixed temperature.

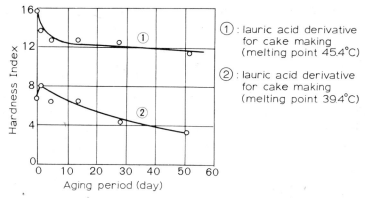

Fig. 3.61.   Effect of aging on the hardness index of fats for cake making (moulded at 10 °C, measured at 25 °C).

The results of these experiments are summarized in Figure 3.62. Some of the samples stored at temperatures above $0\,°C$ showed an exponential increase of the resistance against spread with storage, and reached an equilibrium state about four weeks after the storage. In the case of the samples which were stored at temperatures below $0\,°C$, it was not possible to find any increase of the resistance against spread, so that the spreadability of butter could be regarded as constant during storage in

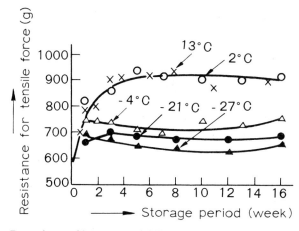

Fig. 3.62.   Dependence of butter spreadability upon temperature and aging period.

the low temperature range. Differences in the mechanical properties of butter appeared in a temperature range from $-4\,°C$ to $2\,°C$, and, therefore, it was suggested that phase transition occurs in the glyceride molecules in the system in this temperature range.

Mohr and Drachenfels [111] have tried to explain the factors affecting the increase in hardness of butter by setting from a morphological viewpoint, and they concluded from the results obtained using a polarizing microscope that the increase of hardness is brought by the formation of a scaffolding structure in the system due to growth of the fat crystals. The crystal nuclei appeared in butter at a relatively low temperature ($6\,°C$), but the rate of crystal growth was pronouncedly slow at this temperature, even though there were very many nuclei in the system. At a relatively high temperature ($19\,°C$), the rate of crystal growth became rapid, although few crystal nuclei appeared in the sample at this temperature. Mohr and Drachenfels also believe that the glassy state can appear in butter when the sample is cooled to an extremely low temperature, because nuclei do not appear, and crystals do not grow, at this temperature. Therefore, both hardness and spreadability can be found in butter, as has been illustrated by Huebner and Thomson [110]. De Man and Wood [112] have evaluated the concentration of solid fat in butter by measuring the specific volume of butter fat by thermal dilatometry, and they obtained results which are in good agreement with those obtained by other investigators.

Van den Tempel [113] has measured torsion creep patterns, using a coaxial type cylinder rheometer, of cylindrically moulded fat which was prepared by dispersing glyceryl tri-stearate in liquid paraffin. The sample showed instantaneous and retarded elasticities in a range of small deformations, and had a high value of the apparent viscosity in large deformation due to the development of macroscopic cracks in the system. Van den Tempel pointed out that the habit of the network structure of fat crystals depends on the density and strength of the linkages between the crystals, and that recoverable weak linkages are due to van der Waal's forces between the crystals, while non-recoverable strong linkages are produced by mixed crystallization of the respective fat components, or by the entanglement of crystals.

Van den Tempel's model should be useful for explaining the rheological properties of edible fats in terms of the habits of the network structure of fat crystals provided that the edible fats contain glycerides whose melting temperatures are relatively high, but it is necessary to do a number of experimental studies to prove this point.

(iv) *Dependence of Viscosity upon Kneading*

As has been previously described in this chapter, edible fats are generally non-Newtonian plastic bodies i.e. the viscosity depends on the shear stress and rate of shear, and a small stress does not cause permanent deformation of the system. Therefore kneading, which is a unit processes involving larger shear stresses than for measuring viscoelasticity, should affect the rheological properties of the sample. In practice, the process of kneading confers a preferable viscosity on butter, margarine and shortening due to homogenization of the inner structure in the system.

Figure 3.63 shows a plot of shear stress against shear rate for kneaded butter at 25 °C, as measured by use of a cone-and-plate viscometer.

Unkneaded butter has a yield value of about 3000 dyn cm^{-2}, and its shear rate increases very slightly with increasing shear stress until the stress reaches about 8000 dyn cm^{-2}, and then the shear rate becomes extremely high while the stress increases. With increasing kneading time, the dependence of shear rate upon the shear stress becomes increasingly pronounced, although the yield value of respective

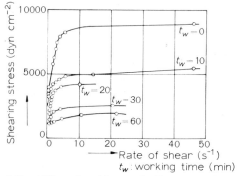

Fig. 3.63.   Effect of working on the flow curve of butter.

samples was kept at about 1000 dyn cm^{-2} even in the case of samples which were kneaded for 60 min. We have a number of reports on the kneading of butter, in which Mulder [114] has tried to repeat periodically the kneading of butter at 15 °C during one week's storage at 10–12 °C, and has obtained results, as shown in Table 3.20. That is, the hardness of butter is lowered by each kneading, and it recovers during storage for one week, although the degree of recoverable hardness is gradually reduced by repeated kneading.

TABLE 3.20

Change of butter hardness due to the working

Treatment	Hardness of butter (g)	
	I	II
Storage for one week at 10 ~ 12 °C	250	430
Reworking	73	190
Storage for one week at 10 ~ 12 °C	210	380
Reworking	70	101
Storage for one week at 10 ~ 12 °C	171	306

Burger and Scott Blair [115] have classified phenomenologically the rheological properties of butter, and have described the time dependence of isothermally recoverable change of viscosity as thixotropy. It is clear from Table 3.20, however, that the behaviour of butter may be described as non-perfectly recoverable thixotropy. Van den Tempel [113] has mentioned that the permanent deformation of gelled fat caused by shear is brought about by the breakdown of linkages between crystals in the network structure, and that recoverable deformation occurs due to the weak interaction between the crystal molecules. Although butter is too complicated a system for its complex properties during deformation to be analyzed, it may be possible to compare the inner structure and properties of butter with those of the model fat system employed by van den Tempel; i.e. the internal structure of fat crystal in butter is broken down by the process of kneading, but the broken crystals gradually recover structure during the aging process. Such behaviour of the crystals can be confirmed by measurement of the phase transition of the system, together with experimental studies on the rheological properties.

It is well known that the coefficient of thermal expansion of crystallized fat is different from that of liquid fat, so that the correlation between the specific volume and crystallinity can be obtained from a plot of the specific volume against temperature. Avrami [116] has made kinetic studies on the crystal growth of metal, and has proposed the following theoretical relation;

$$1 - \frac{v_\infty - v_t}{v_0 - v_\infty} = \exp\left(- k \, t^n\right) \tag{19}$$

where v_0, v_∞, and v_t are the specific volumes of the initial state, equilibrium state, and the state at t hours after the initial state, respectively, k is the rate constant for crystal growth, and n is a constant which depends on the mechanism of crystallization. When the dependence of both change of specific volume and of viscosity upon aging is measured after kneading at 20°C, and if one speculates that the change of specific volume is inversely proportional to that of the viscosity, as follows;

$$1 - \frac{\eta_t - \eta_0}{\eta_\infty - \eta_0} = \exp\left(- k \, t^n\right) \tag{20}$$

it is possible to obtain a relation for the viscosity change similar to that for the specific volume change, as shown in Figure 3.64. The line B in this figure shows the viscosity change of kneaded butter at 20°C as represented by Equation (20), and the slope of the line B is the same as that for the change of specific volume, so that it can be deduced that the crystallization of fat is intimately related to the consistency of the system.

(v) *Effect of Additives*

In butter the water globules are dispersed homogeneously in the fat component. Shortening is prepared by mixing together a few kinds of oils and fats. Chocolate contains condensed milk, lecithin, sugar, etc. Mayonnaise is prepared by dispersing

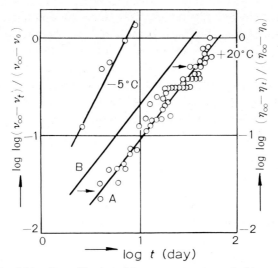

Fig. 3.64. Crystallization of fat components in worked butter.

large quantities of oils and fats in water. The consistency of these fat products is very important to the evaluation of quality.

Fincke and Heinz [117] have tried to obtain information on the correlation between the melt viscosity of plane and milk chocolate and the quantity of lecithin in the sample. The results obtained are summarized in Figure 3.65. Although the apparent

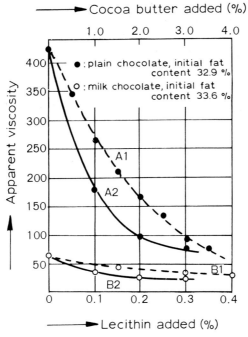

Fig. 3.65. Variation in apparent viscosity with cacao butter and lecithin contents.

viscosity of molten chocolate increases with increasing amount of fat a very small quantity of lecithin plays a big part in the viscosity of chocolate. Chocolate is a dispersed system of sugar, cacao, etc., in cacao fat, and Harbard [118] has suggested that the cacao fat acts not only as the suspending medium but also as the dispersing agent for the dispersed components through adsorption on to the surface of the dispersed phase. Therefore, properties of the dispersion state such as porosity of the dispersed particles in the system play an important role in the viscosity of chocolate, as follows [118];

$$\frac{\eta_{pl}}{\eta_0} = \left(1 - \frac{C}{1 - v}\right)^{-k} \tag{21}$$

where η_{pl} is the plastic viscosity of chocolate, η_0 is the viscosity of free liquid phase, C is the proportion of solids in suspension (by volume), v is the proportion of voids in the packed solids (by volume), and k is a constant. In Equation (21), the term $C/1-v$ corresponds to the dispersion state factor, so that the viscosity of molten cacao fat and the physico-chemical interaction between the cacao fat and the dispersed solids play a very big part in the plastic viscosity of molten chocolate.

Oils are usually hardened by hydrogenation for providing glycerides of saturated fatty acids. The plasticity of shortening can be controlled by the concentration of liquid cotton seed oil in the system, because shortening is prepared by mixing the hydrogenated cotton seed oil together with liquid unhydrogenated oil.

Figure 3.66 shows a plot of the hardness of shortening against the amount of liquid cotton seed oil in the system. It is possible that the crystals of hydrogenated oil construct a macroscopic scaffolding structure in the same way as butter. This belief may be supported by the fact that when shortening is extruded repeatedly by a pump, the hardness of the sample is pronouncedly lowered by the first extrusion, but it is not affected so much by further extrusions, i.e. the macroscopic structure of the crystals in the system is alsmost completely destroyed by the first treatment.

Dolby [120] has studied the effects of various processes on the consistency of butter

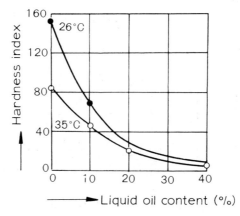

Fig. 3.66.   Hardness of fat blends of solid fat and liquid oil [109].

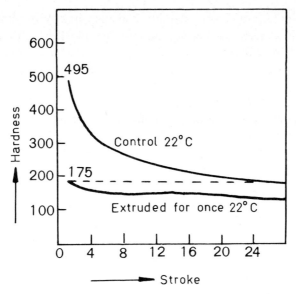

Fig. 3.67.   Extrusion effect of shortening.

e.g. the method of pasteurization, cooling rate after the pasteurization, setting temperature, churning temperature, degree of kneading, etc. He found that the consistency of butter was much influenced by the cooling rate after pasteurization, and that rapid cooling made a quite hard sample.

As shown in Table 3.21, Kapsalis *et al.* [121] have found a correlation between the

TABLE 3.21

Correlation between the spreadability of butter evaluated by consumers and the hardness of butter measured by means of a consistometer

Number of sample	Value of consistometer (g)				Consumer panel	
	Spreadability		Hardness		Spreadability	
	Range	Average	Range	Average	Range	Average
2	206– 403	332	120–144	132	9.5–8.8	9.2
2	574– 578	576	138–158	149	9.0–8.8	9.2
4	610– 690	648	172–197	183	8.1–6.2	7.4
7	724– 789	761	176–230	192	7.4–6.0	6.9
13	831– 896	869	196–292	222	8.4–4.7	7.0
22	903– 991	943	188–273	226	8.4–4.9	6.4
17	1004–1090	1041	200–285	241	7.2–4.5	6.1
16	1102–1183	1137	218–361	268	6.7–4.0	5.8
10	1216–1286	1250	265–430	332	6.4–3.8	5.2
9	1333–1400	1367	292–451	365	6.1–4.5	5.5
5	1426–1440	1435	366–498	448	5.2–4.7	4.9
2	1514–1542	1514	500–500	500	3.4–3.4	3.4
all the samples	206–1542	1017	120–500	258	9.8–3.4	6.2

sensory assessment of butter spreadability and the objectively evaluated hardness of butter in their statistical investigations on the consistency of 109 samples of market butter. The sensory evaluation was made by a points method ranging from eleven to one according to the grade of the spreadability.

3.5.2. CONSISTENCY OF ICE CREAM

Although the major component of ice cream is non-fat milk solids rather than fats, a small quantity of fat strongly affects the texture of ice cream. Ice cream is prepared from ice cream mix. This contains milk, sugar, and stabilizing agent, and it is aged so as to permit hydration of milk protein in the system. After the aging the mix is whipped and cooled in a freezer. The so-called soft ice cream can be obtained by taking the mix out from the freezer at a low temperature. The air content of soft ice cream is generally small.

Ordinary ice cream is extruded from the freezer in a temperature range from $-5\,°C$ to $-8\,°C$, and the air content is nearly 100%, which is usually expressed by the term overrun, i.e. the density of ice cream relative to that of the ice cream mix as the standard. Ordinary ice cream is prepared by refrigeration of soft ice cream in either a cup or a coated tin in a temperature range from $-30\,°C$ to $-20\,°C$, at which temperature ice crystals appear in the bulk of ice cream. Therefore, the consistency of ice cream is much affected by the constitution of the mix, the crystalline state of ice, overrun, etc.

Butter fat, lactose, sugar, and emulsifying and stabilizing agents in ice cream play a significant part in the formation of ice cream texture. Nickerson and Pangborn [122] have investigated the factors affecting the consistency of ice cream. They employed samples which consisted of 12% milk fat and of various quantities of sugar or non-fat milk solids. As shown in Table 3.22, the results indicated that the melting rate in-

TABLE 3.22

Effects of sucrose and milk solid on the rate of melting and hardness of ice cream

	13% Sugar 11% MSNF	15% Sugar 11% MSNF	17% Sugar 11% MSNF	19% Sugar 11% MSNF	15% Sugar 10% MSNF	15% Sugar 12% MSNF
Consistency in centipoise (ice cream mix at 1 °C)	136	143	160	230	143	172
Melting						
5 min	0	0	0	0	0	0.5
10 min	0	2	1.5	2	2	2.5
15 min	2.5	9	9	9	8	9
20 min	10	18.5	21	20	20.5	18.5
25 min	20.5	31	36	32	33	31
30 min	31.5	41	49	47	45	43
Hardness						
$-17\,°C$	12.4	14.4	17.2	17.8	13.1	14.8
$-6\,°C$	23.2	27.0	28.0	37.9	26.2	27

creased with increasing quantity of sugar. The hardness of ice cream at the freezing temperature also increased when the quantity of sugar was increased, but Nickerson and Pangborn [123] have pointed out that both the thermostability and hardness of ice cream are not influenced by small variations in the milk solid content.

The influence of gelatin, when used as a stabilizing agent in ice cream, upon consistency is discussed in a report by Leighton *et al.* [124], in which they made creep measurements. In the case of 100% overrun, the viscosity of ice cream was 1.98×10^{10} centipoise.

Figure 3.68 shows a plot of the viscosity of ice cream against the quantity of milk fat in the system; the viscosity was represented by the value relative to that of a standard sample which contained 3% milk fat. The viscosity of ice cream increased with increasing quantity of fat in a range of fat contents from 0 to 3%, and then the viscosity decreased with further fat addition up to 15–18%, when the system possessed the most preferable body and texture. In general, the increase of fat content promoted a lubrication effect between the ice crystals in ice cream, and it, therefore, induced a decrease in consistency. Although the stabilizing agent shows a similar effect with fat, an excess of stabilizing agent produces a high consistency in ice cream.

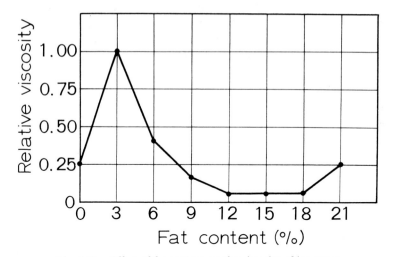

Fig. 3.68.   Effect of fat content on the viscosity of ice cream.

Higher homogenization temperatures induce a greater effectiveness on the part of the stabilizing agent. The viscosity of ice cream is not always useful for judging quality, but it is possible to compare the difference of quality in respective samples using the viscosity data.

Refrigerated condensed milk may be utilized as a model for refrigerated ice cream mix without overrun and sugar. The authors [125] have tried to obtain information on the relationship between rheological properties and the results of chemical, microbiological, and sensory studies using both refrigerated condensed skim milks and whole milk in order to test the possibility of using these samples for reduced milk. Firstly, the refrigerated condensed milk was moulded in a cylinder block, and then the viscoelasticity of the sample was tested by creep measurements, from which the mechanical behaviour of the sample could be represented by use of the three-element mechanical model. The viscosity parameter in the model appeared to be in a range of 5×10^9–2.5×10^{12} poise, although the viscosity of the quick frozen sample was generally lower than that of the slow frozen condensed milk. The quality of the stored condensed milk for five months at $-20\,°C$ was also better in the quick frozen sample than in the slow frozen one.

Figures 3.69a, b and c show the creep patterns for samples of both refrigerated condensed milk and ice cream. Although it was predicted that the deformation of these samples would be brought about by either plasticity or energy elasticity, because of the existence of a large quantity of ice crystals in the systems, the creep patterns obtained obviously depended on the temperature, so that entropy elasticity might play a big part in the deformation rather than energy elasticity. However, with decreasing quantity of milk solid in the refrigerated condensed milk the temperature dependence of the creep pattern was gradually lost, and therefore, the deformation perhaps became more energy elastic. Such transition might be brought about by a decrease of elastic contribution by the protein molecules as the amount of ice crystals increased. This tendency could also be seen in the case of ice cream studied at a lower temperature range. Shaw [126] has obtained both patterns of creep and recovery with ice cream, from which the elastic properties of ice cream were explained by means of a network structure of protein molecules. It seems, however,

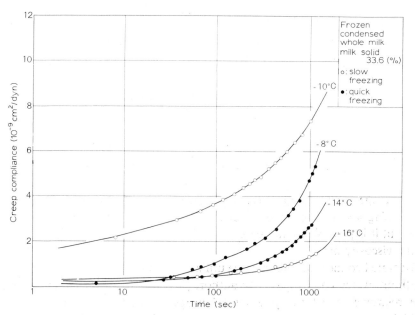

Fig. 3.69a.   Creep curve of frozen condensed whole milk.

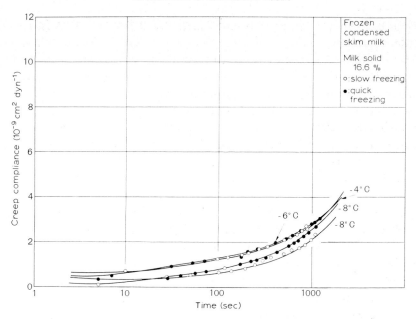

Fig. 3.69b.   Creep curve of frozen condensed skim milk.

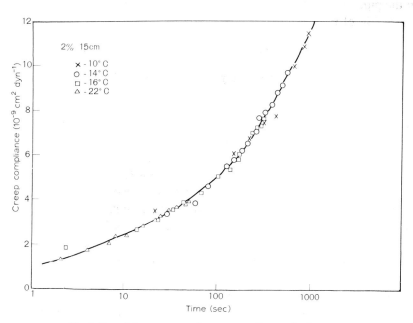

Fig. 3.69c.   Master curve of creep compliances for ice cream.

that the air cells may play an important role in the elastic behaviour of ice cream.

In order to obtain preferable texture in ice cream, it is necessary to homogenize the milk fat into fine globules $(1–2\,\mu)$. Meanwhile, the essential thing is to prevent coagulation between the fat globules, because the whipping process induces coagulation and coalescence of the globules. Therefore, an emulsifying agent is introduced into the system for lowering the interfacial tension between the fat phase and the suspending fluid.

It is desirable that the surface denaturation of protein molecules on the surface of the fat globules produces a mechanism for preventing coalescence between the fat globules. Emulsions generally show a tendency for irreversible coagulation between the dispersed liquid droplets. Therefore, the rate of coagulation must be reduced if a preferable texture is to be obtained in ice cream. Carefully prepared ice cream mix consists of stable and fine fat globules, and such a system is also able to stabilize air cells. As far as solubility of the emulsifying agent in the fat phase is concerned, Stistrup and Andreasen [127] have reported that the emulsifying agent which contains a large amount of unsaturated fatty acid is more soluble than that prepared from saturated fatty acid at a relatively low temperature, and that the protective layer on the surface of the fat globules consists in large part of glyceryl mono-unsaturated fatty acid esters. Actually, the protective layer is mechanically so strong that it can bestow protective action against fracture, which is caused by refrigeration of fat components and by the coagulation of fat globules. The dispersion state of ice cream is generally affected by the conditions of each manufacturing process. For example, when ice cream is extruded from a freezer, the system shows plug flow in the pipe, so that the ice cream is only sheared near the pipe wall. Therefore, the coagulation of fat globules and the break down of air cells also only occurs in that part of the system near the pipe wall.

3.6. Consistency of Powdered Foods

If foods can be transformed into a powdered state, we are able to provide many advantages for foods such as desiccation, storage, restoration, mixing, transport, etc., although these special features much depend upon the means used to transform foods into powder.

There are many techniques for reducing foods to powder. For example, mechanical grinding methods for solid state foods (wheat flour, powdered cheese, cocoa, etc.), spraying desiccation for liquid state foods (powdered milk, powdered juice, instant coffee, etc.), crystallization or precipitation method (sugar, salt, potato starch granule, etc.). It is, however, believed that the consistency of powdered foods is influenced by the components of the food, apart from sugar or salt because of the simplicity in their components, and by the powdering process.

Recently, the consistency of powdered foods has also been concerned with many important subject matters in the field of industry such as transport, packing, restoration, etc.

3.6.1. VARIATION OF DENSITY OF POWDERED FOODS

Table 3.23 shows remarks about the packing properties of various powdered foods such as the porosity in a closely packed state ε_{max}, porosity in a loosely packed state ε_{min}, and the ratio of decreasing volume to the orginal volume $\Delta V/V (\%)$. Although the ratio of the ideal porosity in loosely packed spheres ε_t to $\varepsilon_{max}(e_1)$ should become greater than unity, most powdered foods actually have a smaller value than unity. The ratio $\varepsilon_t/\varepsilon_{min} = e_2$ also approaches unity for ideal spheres, so that it is possible to deduce from the table that starch granules and both powdered products of whole milk and skim milk may be nearly spherical, while non-spherical powdered foods are instant skim milk, wheat flour, and crystals of both lactose and sugar. It is possible that $e_1 < 1$ may be brought about by the formation of a so-called 'bridge' between the powder particles. The formation of the bridge also induces heavy flow of the powder. Powdered foods generally show this phenomenon.

TABLE 3.23
Properties for packing of various powdered foods [127]

Sample	Mean particle diameter (μ)	Density Loose	Density Close	(a) $\Delta V/V$ (%)	(b) ε_{max}	(c) ε_{min}	(d) e_1	(e) e_2	(f) Density
Wheat flour	120	0.484	0.606	20	65.2	56.7	0.730	0.460	1.400
Lactose	100	0.586	0.812	28	61.4	46.5	0.775	0.560	1.520
Sugar	400	0.660	0.880	25	58.0	44.0	0.820	0.590	1.578
Powdered whole milk	110	0.589	0.710	17	43.5	31.6	1.100	0.824	1.039
Powdered skim milk	105	0.589	0.746	21	51.7	38.9	0.920	0.667	1.224
Skim milk for instant use	300	0.284	0.312	9	76.8	74.4	0.620	0.351	1.224
Starch (soluble)	55	0.810	0.966	16	42.0	31.0	1.130	0.840	1.400

(a) Decreasing ratio of powder volume (b) Porosity in a state of close packing (c) Porosity in a state of loose packing (d) $\varepsilon_t/\varepsilon_{max}$, $\varepsilon_t = 47.64$ (e) $\varepsilon_t'/\varepsilon_{min}$, $\varepsilon_t' = 25.95$ (f) In the case of both powders of whole milk and skim milk, one may generally observe an air content of about $10 \sim 13\%$.

3.6.2. COEFFICIENT OF INTERNAL FRICTION OF POWDERED FOODS

Measurement of the coefficient of internal friction can be made with a couple of cylindrical vessels, which are placed one upon another. When the lower vessel is made to slide in a fixed direction after introducing the sample into the vessels, a stress develops in the upper vessel. In the case of powdered foods, the shearing stress increases with increasing shear deformation. Meanwhile one may observe a pulse pattern in the increase of shearing stress. This is a special feature of powdered foods, and it is brought about by the complicated structure on the surface of the powder particles.

Figure 3.70 shows a plot of the shearing stress against the vertical pressure in the measuring vessels for various powdered foods. These linear relations in Figure 3.70 satisfy Coulomb's empirical rule, and each slope corresponds to the coefficient of internal friction of each sample. The coefficients for the various powdered foods are distributed in a range of 0.2–0.7, and with increasing coefficient the shearing stress increases, while the sample shows heavy flow. The intercept at the ordinate in Figure 3.70 means that the powder particles can not be moved by the frictional resistance between the particles without any vertical pressure. This resistance appears to be brought about by the adhesion between the powder particles, and it is conspicuous in both powdered products of ice cream mix and of whole milk, which contain a relatively large amount of fat.

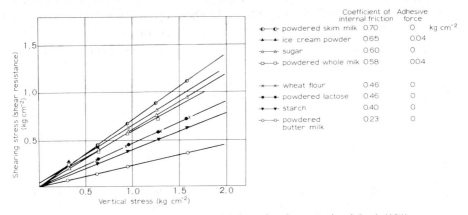

Fig. 3.70.   Coefficient of internal friction of various powdered foods [129].

Other factors affecting the coefficient of internal friction are the porosity and shear rate. With increasing porosity the coefficient decreases, while the shear rate becomes independent of the coefficient. For small porosity, however, the coefficient of internal friction increases through a minimum value with increasing shear rate.

3.6.3. TAPPING OF POWDERED FOODS

When powder is packed in a tin or bag, we are accustomed to tapping the powder in the container. In the case of spontaneous flow, the powder is packed so loosely that the porosity will take a maximal value. Therefore, tapping is necessary to reduce the porosity for close packing of the powder particles. Tapping is also useful for estimating the consistency of powdered foods.

As has been described in Chapter 2 of this book, there is a relationship between the number of tappings N and the volume strain γ such that $\gamma = abN/(1+bN)$ so that a plot of N/γ against N is linear. Figures 3.71a and b show plots of N/γ against N for various powdered foods.

If the minimum volume of powder after infinite tapping is v_∞, the volume strain can be given by $v_0 - v_\infty/v_0$, where v_0 is the volume in the most loosely packed state of powder. When N approaches ∞ in the relation $\gamma = abN/(1+bN)$, γ also ap-

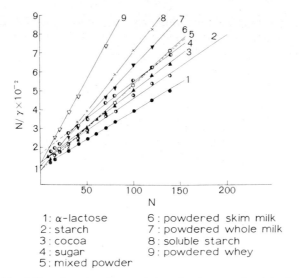

1: α-lactose
2: starch
3: cocoa
4: sugar
5: mixed powder

6: powdered skim milk
7: powdered whole milk
8: soluble starch
9: powdered whey

Fig. 3.71a. Tapping characteristics of various powdered foods.

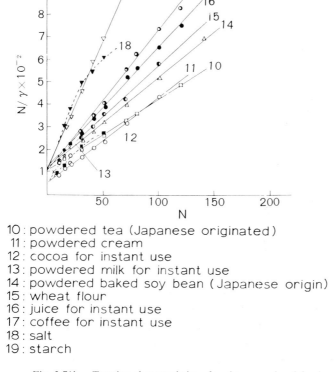

10: powdered tea (Japanese originated)
11: powdered cream
12: cocoa for instant use
13: powdered milk for instant use
14: powdered baked soy bean (Japanese origin)
15: wheat flour
16: juice for instant use
17: coffee for instant use
18: salt
19: starch

Fig. 3.71b. Tapping characteristics of various powdered foods.

proaches a, so that the physical meaning of a should correspond to infinite compressibility of the powder at $N \to \infty$. Therefore,

$$b = \frac{1}{N} \frac{v_0 - v}{v - v_\infty}. \tag{22}$$

Suppose that the number of tapping N_τ induces the volume $v_h = (v_0 - v_\infty)/2$, then it follows that $N_\tau = 1/b$, which may represent the degree of difficulty for packing the powder by tapping. The terms a, b, and internal friction angle $\tan a$ for various powdered foods are summarized in Table 3.24. One may find in this table that little

TABLE 3.24

Coefficient of internal friction and tapping characteristics for various powdered foods [129]

Sample	Value of a		$b = 1/N\tau$	$\tan a$
	Calcu.	Obs.		
	$\times 10^{-1}$	$\times 10^{-1}$	$\times 10^{-2}$	
Lactose	3.3	2.8	3.7	0.46
Starch	2.9	2.9	3.9	0.40
Cocoa	2.6	2.5	3.2	–
Sugar	2.6	2.5	2.3	0.60
Ice cream powder	2.2	2.1	6.0	0.65
Powdered skim milk	2.2	2.1	6.2	0.70
Powdered whole milk	2.1	1.7	4.8	0.58
Souluble starch	1.6	1.6	8.3	–
Whey powder	1.1	1.0	9.6	–
Wheat flour	2.5	2.0	3.4	0.46
Powdered tea	3.2	2.7	2.9	–
Powdered cream	2.7	2.3	6.8	–
Cocoa for instant use	2.2	1.4	9.1	–
Powdered skim milk for instant use	1.3	0.9	17.5	–
Soybean flour	2.5	2.0	2.3	–
Juice for instant use	1.8	1.6	4.9	–
Coffee for instant use	1.6	1.6	6.0	–
Salt	1.9	0.8	5.2	–
Potato starch	0.9	0.9	11.4	–
Powdered butter milk	–	–	–	0.23

volume change appears in whey powder, powdered skim milk for instant use, salt, potato starch granule, etc. by tapping, because the bulkiness (apparent density) in the loosely packed state at an initial stage is relatively large in these samples. This phenomenon is related to the porosity of powder and the interaction between powder particles, and, therefore, the term a acts in opposition to the term b. Moisture affects both a and b for powdered foods, and this tendency is pronounced for powdered milk products because of the complicated structure on the surface of the powder particles. In practice, canning processes or rate of flow from a hopper is also influenced by the moisture content of the sample.

3.6.4. CREEP OF POWDERED FOODS

Although powder particles are solid, powder as an assembly of solid particles shows flow properties depending on the change of porosity. Taneya and Sone [131] have studied the creep behaviour of powdered foods using a compression technique, and they have confirmed that the flow properties of the samples can be represented by use of Nutting's equation $\gamma = \phi^{-1}\sigma^{\beta}t^{k}$, where γ is the strain, ϕ is the firmness, σ is the stress, t is the time, and β and k are constants.

Table 3.25 shows the constants β and k for powdered skim milk, powdered whole milk, sugar, cocoa, and starch granule. The data in the table show that the term β takes various values with different powdered foods, although the term k does not show so much change for different foods, excepting for starch granules where k is zero. When $k=0$ it means that the sample deforms instantaneously. In the case of powdered skim milk, the term β initially decreases with increasing moisture content of the sample, but then it increases when the moisture content exceeds 10%, but the term k is not affected by the moisture content. Also with decreasing size of powder particle, k increases while β decreases. One may expect to exert the same effect on the flow properties of powder either by increasing the moisture content or by decreasing the particle size.

TABLE 3.25
Values of constants in Nutting equation

Sample				k	β	$\text{Log}\,\phi$
Powdered skim milk	Moisture		7.8 (%)[a]	0.25	0.99	7.96
			10.6	0.25	0.65	5.73
			14.3	0.24	0.73	5.69
	Granule size	590	297 (μ)[b]	0.19	0.82	6.60
		297	125	0.26	0.64	5.73
		62	53	0.35	0.33	4.22
Powdered skim milk				0.25	0.90	5.60
Powdered whole milk				0.15	1.90	15.43
Sugar				0.26	1.56	13.22
Cocoa				0.20	0.95	9.18
Starch granule				0.00	0.64	4.68

[a] Granule diameter: $297 \sim 125\,\mu$
[b] Moisture: 10.6%

3.6.5. KINEMATIC PROPERTIES OF POWDERED FOODS

It has been confirmed by Matheson *et al.* [132] that powders generally show Newtonian flow in the relationship between the speed of air current and the torque of the system. Benarie [133] has measured the strength of adhesion between oil-coated sands using a Couette type viscometer. Ono *et al.* have reported that when the flow properties of powder are measured by means of a rotational viscometer, a little fluctuation can be detected in the torque, and that the torque is not proportional to the rest angle of the sample. Ono and Taneya [134] have tried to obtain information on the relationship between the rotational speed and the torque with various powdered foods using a Green type rotational viscometer, and they have observed that the fluctuation in the torque is most pronounced in the initial period of rotation. The fluctuation pattern may be related to the adhesiveness of the powder particles, i.e. the larger the adhesiveness the smaller the amplitude of fluctuation, or the dryer the powder, the larger the amplitude of fluctuation. The relationship between the number of rotations N and the torque can be expressed by the equation,

$$\log |(T - T_\infty)/(T_\infty - T_0)| = - kN \tag{23}$$

where T_0, T_∞ and T are the torques of initial rotation, infinite rotation, and increased torque with increasing number of rotation, respectively, and k is a constant which corresponds to the rate of packing. The term k is also related to the density change on tapping.

Figure 3.72 shows a plot of the constant k against the number of rotations, in which one may find that with increasing number of rotations the term k becomes constant for various samples, although the k value depends on the kind of powder in the quasi-static state. The rate of packing is not affected by the shape and composition of the powder in the dynamic region.

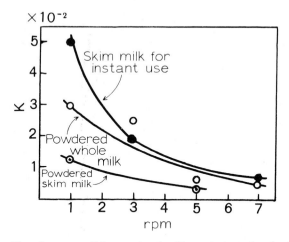

Fig. 3.72.   Plot of constant K for powdered milks against rotational velocity [135].

It is difficult to measure the rest angle of powdered foods using the ordinary method (the rest angle of powder is usually represented by the angle subtended on the surface by the powder accumulated on the plane), because the particles of powdered food so frequently interlink with each other that the particles do not always accumulate as a circular cone. Taneya *et al.* have tried to measure the rest angle of powdered foods by mixing the sample with glass spheres in various ratios and extrapolating the mixing ratios of glass spheres to zero, and they have found that cocoa, sugar and powdered whole milk show a rest angle about 90°. These systems are not dry, so that to measure the precise mass of the system is difficult, because the powder particles readily form lumps. In general, the rest angle becomes increasingly stable with increasing size of powder particles. Table 3.26 shows apparent density and rest angle data for various grains and granules. The rest angle of these materials are relatively smaller (20–30°) than those of powdered foods, and these grains are very fluid. Finely powdered foods do not flow easily because fine particles induce various phenomena such as adhesiveness, complicated shape and surface structure, electrostatic interaction between the particles due to the friction, etc. The correlation between the consistency of powdered foods and the electric charge on the surface of the particles is not fully investigated yet, although it is widely recognized that the adhesion of powdered milk to the inside wall of the drying tower during the desiccation of milk may be brought about by the electric charge on the particles.

TABLE 3.26

Rest angle of various grains and granules

Sample	Apparent density (g cm^{-3})	Rest angle (ϕ^0)
Rice	0.80	20
Powdered rock salt	1.20	20
Salt granule	1.30	31
Soybean	0.64	27
Wheat	0.77	23
Coffee bean	0.67	25

3.6.6. SOLIDIFICATION OF POWDERED FOODS

It is well known that medical supplies are frequently modified to tablets. Some powdered foods are also processed as tablets in order to provide a more convenient form for use. For example, we can get the so-called crystal milk, solid soup, carmine, cube sugar, etc., while attempts have been made to solidify refreshing drinks. It is possible to quote many works on the compression of powder such as Athy [136], Walker [137], Kawakita [138], Taneya and Takada [139], etc. As far as solidification of powdered foods is concerned, however, we can only refer to an investigation reported by Chuchlowa [140]. The techniques for solidifying powdered foods have been developed by the respective manufacturers of the products sold on the market. Such techniques can generally be classified in two ways. First, the-so-called dry method, in which the powder is processed to tablets with or without binding agent; the second, is the so-called wet method, in which the powder is provided with some quantity of water for making tablets. In the latter case, however, powdered foods are so often contaminated with water that the dry method is widely employed for preparing solidified foods excepting recrystallised sugar.

Taneya et al. have tried to solidify various powdered foods and these mixtures using the dry method, and they have measured the mechanical strength, brittleness, water absorbability of the tablets and the quantity of liquid isolated from the sample due to solidification. Figures 3.73 and 3.74 show the relationship between the apparent density and the compressive pressure for various powdered foods.

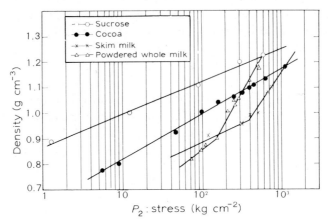

Fig. 3.73.   Diagram of density and pressure in the process of solidification of powdered foods.

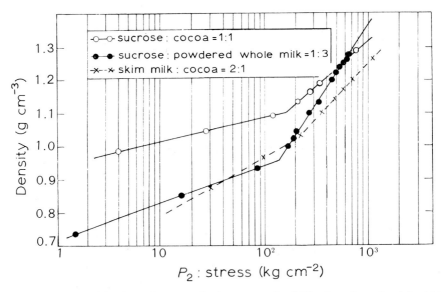

Fig. 3.74.   Diagram of density and pressure in the process of solidification of powdered food blends.

In the case of uniphase systems such as sugar, cocoa, etc., the plot of the apparent density against pressure became linear, but the slope of the same plot for powdered milk or powdered food blends suddenly changed at a certain pressure, which depended on the nature of the sample. This change in slope appeared to be brought about by the breakdown of structure in the powder particles due to the pressure.

When powdered foods are processed into tablet form with a tablet machine, the stress developing in the lower mould of the machine is generally about 20–30% of that appearing in the upper mould, although the damped value of the stress from the upper part to the lower part is about 60% in some cases. The mechanical strength of tablets generally increases with increasing compressive pressure. In the case of powdered whole milk, however, the mechanical strength reaches an equilibrium above a pressure of about 500 kg cm^{-2}. Whole milk contains such large quantities of fat that the fat is isolated from the system in proportion to the degree of pressure, but the quantity of isolated fat does not increase above a pressure of about 500 kg cm^{-2}.

It is possible to conclude from the above facts that the mechanical properties of food tablets are much influenced by the method of mixing, kind of food, presence of fat, etc.

3.6.7. FLOWING-OUT PROPERTIES OF POWDERED FOODS

When powder is removed from a hopper or orifice, the powder shows very complicated flow patterns as shown in Figure 3.75. The pattern (E) shows the layer at rest, (B) is a slow moving layer, (A) corresponds to the more rapid moving layer than (B), and (C) is a very rapid moving layer. The intersecting face in both patterns (A) and (C) reduces gradually in proportion to the reciprocal square of the sample height in the hopper. The layer (D) consists of the free falling particles, and the diameter of the powder flux is generally smaller than that of the orifice.

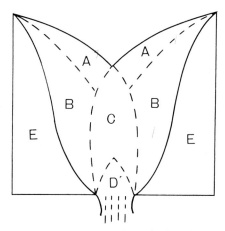

Fig. 3.75.   Flow pattern of powder from hopper.

There are some empirical equations for expressing the flow rate of powder from a hopper taking account of various factors such as diameter of orifice D_B cm, apparent density of powder ϱ_a g cm^{-3}, density of powder ϱ g cm^{-3}, inclined angle of hopper $\theta°$, diameter of powder particle D_p cm, coefficient of internal friction μ, rate of flow W, and shape factor K, as follows:

Tanaka and Kawai's formula [141]

$$W = 5.634 \, \varrho_a D_p^{2.5} \, (D_p/D_B) \, \mu^{-0.32} \, (\tan \theta/2)^{-0.32} \tag{24}$$

in $D_p/D_B < 0.1$

$$W = 3.13 \, \varrho_a D_p^{2.5} \, (D_p/D_B)^{-3.0} \, \mu^{0.32} \, (\tan \theta/2)^{-0.32} \tag{25}$$

in a range of $0.1 < D_p/D_B < 0.23$.

Raush's formula [142]

$$W = 5.04 \, \varrho_a D_B^{2.745} \, D_p^{-0.245} \, (CC_0/\sqrt{\mu}) \tag{26}$$

where C and C_0 are the correction factors for interrelationship between the hopper wall and D_p/D_B, and the hopper angle and D_p/D_B, respectively.

Beverloo and Lenigar's formula [143]

$$W = 18.25 \, \varrho_a (D_B^{-1.4} D_p)^{2.5}. \tag{27}$$

Flowler and Glastonbury's formula [144]

$$W = 8.2 \, \varrho_a D_B^{2.685} \, (KD_p)^{-0.185}. \tag{28}$$

Kawai has examined experimentally the accuracy of the empirical equations listed above with various powdered foods, as shown in Table 3.27. It must be recognized that the experimental results do not really agree with the calculated values obtained by use of these equations.

TABLE 3.27

Application of empirical equations to the rate of flow of powdered foods from hopper

Investigator	Sample	D_p cm	D_B cm	ϱ g cm^{-3}	ϱ_a g cm^{-3}	μ	k	W_{obs} g s^{-1}	$\dfrac{W_{obs} - W_{cal}}{W_{obs}} \times 100$
1. Tanaka	Millet	0.123	1.19	1.20	0.69	0.70	–	10.5	+ 6.0
Kawai	Sugar crystal	0.130	0.90	1.59	0.82	1.06	–	4.45	+ 15.0
2. Rausch	Soybean	0.760	3.81	1.16	0.76	0.81	–	145.3	– 1.0
3. Beverloo	Rapeseed	0.170	2.50	1.12	0.67	0.54	0.91	97.0	– 2.3
and	Turnipseed	0.170	2.50	1.12	0.68	0.483	0.79	100.0	– 4.9
Leniger	Linseed	0.25	2.50	1.16	0.69	0.69	0.48	75.0	+ 14.0
	Spinageseed	0.35	2.50	1.19	0.57	0.61	0.54	63.1	– 2.2
4. Flowler	Sugar	0.109	2.131	1.59	0.908	0.754	0.65	120.0	– 23.0
and	Rapeseed	0.192	1.68	1.103	0.709	0.354	1.00	38.5	– 17.6
Glaston-	Rice	0.315	2.53	1.432	0.882	0.664	0.893	93.8	+ 18.0
bury	Wheat	0.413	2.53	1.38	0.86	0.466	0.80	74.0	+ 41.0

3.7 Consistency of Apparently High-Elastic Foods

Sponge cake, marshmallow, bread, etc. consist of so many air cells or bubbles that they show high compressibility and recovery, as shown in Figure 3.76.

Agar gel and pectin jelly have an elasticity below 10^7 dyn cm^{-2}, and they also show easy deformation and high compressibility. Such materials are described as highly elastic bodies.

The apparent elasticity of sponge-like foods is extremely low, even though the walls of the air cells are composed of rigid material. Moreover, the elasticity of foodstuffs

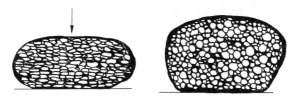

Fig. 3.76.   Schematic tissue of highly elastic food.

is generally low, so that one may obtain very soft foods by blowing bubbles into the systems. Suppose that the diameter of an air cell is R, and that the thickness of the wall of the air cell is t, then the correlation between the inside pressure of the air cell P and the resultant stress S_x can be given by

$$S_x = PR/2t. \tag{29}$$

Equation (29) suggests that with decreasing thickness of the wall of air cells (in other words, with decreasing air cell size at a fixed air content), the system becomes increasingly rigid. Cream is a viscous liquid material and can be transformed into an elastic body by whipping; whipped cream is used to use for decorating cakes.

It is possible that the major component of highly elastic foods are the walls of the air cells, and these should play a big part in the physical properties of sponge-like

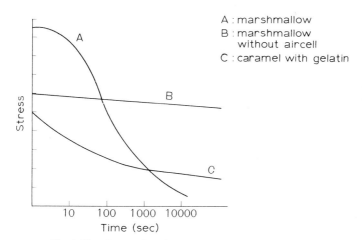

Fig. 3.77.   Stress relaxation curve of marshmallow [145].

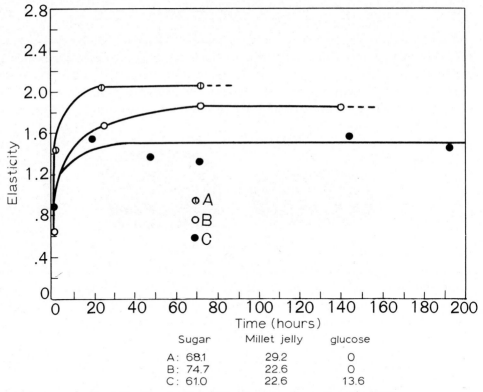

	Sugar	Millet jelly	glucose
A:	68.1	29.2	0
B:	74.7	22.6	0
C:	61.0	22.6	13.6

Fig. 3.78.   Change of elasticity of marshmallow with setting [146].

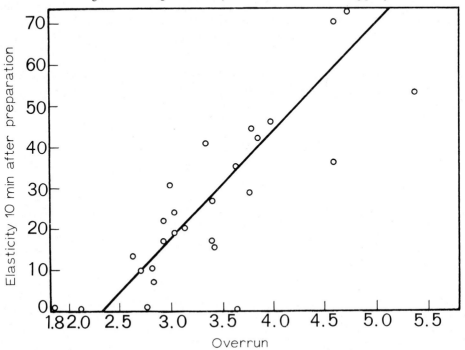

Fig. 3.79.   Overrun-elasticity diagram of marshmallow [146].

foods. Marshmallow is prepared by whipping gelatin, millet jelly, sugar, and egg white so that bubbles are dispersed homogeneously in the system, and the mixture is then solidified in a mould. Figure 3.77 shows the stress relaxation curves of caramel, marshmallow, and non-bubbled marshmallow, respectively. The stress relaxation of caramel can be represented by a simple Maxwell model, but in the case of marshmallow, the relaxation behaviour is much influenced by the presence of bubbles, and the rate of relaxation is rapid for bubbled marshmallow. The elasticity of marshmallow increases with increasing ageing period within 24 h, and then it reaches equilibrium, as shown in Figure 3.78. Factors affecting the setting of marshmallow are temperature, overrun, quantity of gelatin and moisture. An interesting point about these factors is that the plot of elasticity against overrun is represented by a linear relationship during the initial period of setting, as shown in Figure 3.79.

References

[1] Jenness, R. and Palmer, L. S.: 1945, *J. Dairy Sci.* **28**, 653.
[2] Leviton, A. and Leighton, A.: 1929, *J. Phys. Chem.* **33**, 1485; 1932, *ibid.* **36**, 523.
[3] Cox, C. P., Hosking, Z. D., and Posener, L. W.: 1955, *J. Dairy Res.* **26**, 182.
[4] Caffyn, J. E.: 1951, *J. Dairy Res.* **18**, 95.
[5] Glasstone, S., Laidler, K. J., and Eyring, H.: 1941, *The Theory of Rate Processes*, McGraw-Hill, New York.
[6] Whitnah, C. H. *et al.*: 1956, *J. Dairy Sci.* **39**, 356.
[7] Bateman, G. M. and Sharp, R. F.: 1928, *J. Agr. Ress.* **36**, 647.
[8] Puri, B. R., Parkash, S., and Totaja, K. K.: 1963, *Indian J. Dairy Sci.* **17**, 181.
[9] Oppen, F. C. and Schuette, H. A.: 1939, *Ind. Eng. Chem.* (Anal. ed.) **11**, 130.
[10] Munro, J. A.: 1943, *J. Econ. Entomol.* **36**, 769.
[11] Pryce-Jones, J.: 1941, *J. Sci. Instr.* **16**, 39.
[12] Pryce-Jones, J.: 1953, in G. W. Scott Blair (ed.), *Foodstuffs; Their Plasticity, Fluidity and Consistency*, Interscience Publishers, New York.
[13] Davis, P. R. and Prince, R. N.: 1955, *Adv. Chem.* Ser. No. **12**, 35.
[14] Harper, J. C.: 1960, *Food Technol.* **14**, 557.
[15] Whittenberger, R. T. and Nutting, G. C.: 1957, *Food Technol.* **11**, 99.
[16] Hand, D. B., Meyer, J. C., Ransford, J. R., Hening, J. C., and Whittenberger, R. T.: 1955, *Food Technol.* **9**, 228.
[17] Ingram, M.: 1964, Abst. from Z. Berk; *Food Technol.* **18**, 1811.
[18] Ezell, C. H.: 1959, *Food Technol.* **13**, 9.
[19] Charm, S. E.: 1960, *Food Res.* **25**, 351.
[20] Berk, Z.: 1964, *Food Technol.* **18**, 1811.
[21] Charm, S. E. and Merrill, E. W.: 1959, *Food Res.* **24**, 319.
[22] Smit, C. J. B. and Nortze, B. K.: 1958, *Food Technol.* **12**, 356.
[23] Nakagawa, T.: 1951, *J. Chem. Soc. Japan* (in Japanese) **72**, 390, 626.
[24] Greenwood, C. T. and Thomson, J.: 1962, *J. Am. Chem. Soc.* **84**, 222.
[25] Greenwood, C. T.: 1964, *Food Technol.* **18**, 732.
[26] Matz, S. A.: 1962, *Food Texture* **59**, Avi. Pub. Co., Inc.
[27] I.F.T. Committee: 1959, *Food Technol.* **13**, 496.
[28] Olsen, A. G.: 1934, *J. Phys. Chem.* **38**, 919.
[29] Olsen, A. G., Stuewer, R. F., Fehlberg, E. R., and Beach, H. M.: 1939, *Ind. Eng. Chem.* **31**, 1015.
[30] Cheftel, H. and Mocquard, J.: 1947, *J. Soc. Chem. Ind.* **66**, 297.
[31] Owens, H. S., Swenson, H. A., and Schultz, T. S.: 1954, *Adv. Chem.* Ser. No. 11, 10.
[32] Doesburg, J. J. and Grevers, G.: 1960, *Food Res.* **25**, 634.
[33] Matz, S. A.: 1962, *Food Texture*, Avi. Pub. Co., Inc., pp. 64–65.
[34] Ferry, J. D.: 1948, *Adv. Protein Chem.* **4**, 45.

[35] Ward, A. G. and Saunders, P. R.: 1956, in E. R. Eirich (ed.), 'Rheology of Gelatin', in *Rheology, Theory and Applications*, Academic Press, New York.
[36] Leick, A.: 1904, *Ann. Physik* **14**, 139.
[37] Hatschek, E. and Jane, R. S.: 1926, *Kolloid-Z.* **39**, 300.
[38] Sheppard, S. E. and Sweet, S. S.: 1921, *J. Am. Chem. Soc.* **43**, 545.
[39] Kinkel, E. and Sauer, E.: 1925, *Z. Angew. Chem.* **38**, 413.
[40] Cumper, C. W. N. and Alexander, A. E.: 1952, *Australian J. Sci. Res.* **A5**, 153.
[41] Saunders, P. R. and Ward, A. G.: 1954, *Proc. and Intern. Congr. Rheology*, p. 284.
[42] Ferry, J. D.: 1948, *J. Am. Chem. Soc.* **70**, 2244.
[43] Scott Blair, G. W. and Oosthuizen, J. C.: 1961, *J. Dairy Res.* **28**, 165.
[44] Hostettler, H. and Ruegger, H. R.: 1950, *Landw. Jb. Schweiz* **64**, 669.
[45] Scott Blair, G. W. and Oosthuizen, J. C.: 1962, *J. Dairy Res.* **29**, 37.
[46] Berridge, N. J.: 1952, *Analyst* **77**, 57.
[47] Scott Blair, G. W. and Oosthuizen, J. C.: 1962, *J. Dairy Res.* **29**, 47.
[48] Berridge, N. J.: 1942, *Nature* **149**, 194.
[49] Scott Blair, G. W. and Burnett, J.: *J. Dairy Res.* **25** (1958), 297; **25** (1958), 457; **26** (1959), 58; **26** (1959), 144.
[50] Shimizu, H. and Shimizu, W.: 1954, *J. Jap. Soc. Sci. Fish.* (in Japanese) **19**, 596.
[51] Kishimoto, A. and Maekawa, E.: 1962, *Bull. Jap. Soc. Sci. Fish.* **28**, 803.
[52] Shimizu, H.: 1958, *Mem. Coll. Agr. Kyoto Univ.* **68**, June.
[53] Matsumoto, J. and Arai, T.: 1952, *J. Jap. Soc. Sci. Fish.* (in Japanese) **17**, 377.
[54] Shimizu, H. and Shimizu, W.: 1955, *J. Jap. Soc. Sci. Fish.* (in Japanese) **21**, 501.
[55] Kishimoto, A.: 1965, *J. Soc. Material Sci. Japan* (in Japanese) **14**, 264.
[56] Kishimoto, A. and Hirata, S.: 1963, *Bull. Jap. Soc. Sci. Fish.* **29**, 146.
[57] Van Holde, K. E. and Williams, J. W.: 1953, *J. Polymer Sci.* **11**, 243.
[58] Werner, G., Meschtor, E. E., Lacey, H., and Kramer, A.: 1963, *Food Technol.* **17**, 81.
[59] Lee, F. A., Whitcombe, J., and Hening, J. C.: 1954, *Food Technol.* **8**, 126.
[60] Decker, R. W., Yeatman, J. N., Kramer, A., and Sidwell, A. P.: 1957, *Food Technol.* **11**, 343.
[61] Kramer, A.: 1951, *Food Technol.* **5**, 264.
[62] Binder, L. J. and Rockland, L. B.: 1964, *Food Technol.* **18**, 1071.
[63] Deshpande, P. B. and Salunke, D. K.: 1964, *Food Technol.* **18**, 1195.
[64] Breke, J. E. and Sandomire, M. M.: 1961, *Food Technol.* **15**, 335.
[65] Shallenberger, R. S. and Moyer, J. C.: 1961, *J. Agr. Food Chem.* **9**, 137.
[66] Hills, C. H., Whittenberger, R. T., Robertson, W. F., and Cose, W. H.: 1953, *Food Technol.* **7**, 32.
[67] Strohmaier, L. H.: 1953, *Food Technol.* **7**, 469.
[68] Mitchell, R. S., Casimir, D. J., and Lynch, L. J.: 1961, *Food Technol.* **15**, 415.
[69] Tressler, D. K., Birdeye, C., and Muray, W. T.: 1932, *Ind. Eng. Chem.* **24**, 242.
[70] Child, A. M.: 1934, *J. Agri. Res.* **48**, 1127.
[71] Paul, P. C., Bean, M., and Bratzler, L. J.: 1956, *Mich. Agr. Expt. Sta. Tech. Bull.*, 256.
[72] Cover, S., King, G. T., and Butler, O. D.: 1958, *Texas Agr. Expt. Sta. Bull.*, 889.
[73] Tanner, B., Clark, N. G., and Hankins, O. B.: 1943, *J. Agr. Res.* **66**, 403.
[74] Kropf, D. H. and Graf, R. L.: 1959, *Food Technol.* **13**, 492.
[75] Miyada, D. S. and Tappel, A. J.: 1956, *Food Technol.* **10**, 142.
[76] Sale, A. J. H.: 1960, *Soc. Chem. Ind. Monograph* **7**, 103.
[77] Bockian, A. H., Anglemier, A. F., and Sather, L. A.: 1958, *Food Technol.* **12**, 483.
[78] Simone, M., Carroll, F., and Chichester, C. O.: 1959, *Food Technol.* **13**, 337.
[79] Sperring, D. D., Platt, W. T., and Hiner, R. L.: 1959, *Food Technol.* **13**, 155.
[80] Tanner, B. Clark, N. G., and Hankins, O. G.: 1943, *J. Agr. Res.* **66**, 403.
[81] Tuomy, J. M. and Lechnir, R. J.: 1964, *Food Technol.* **18**, 219.
[82] Ritchey, S. J., Cover, S., and Hostettler, R. L.: 1963, *Food Technol.* **17**, 194.
[83] Briskey, E. J., Sayre, R. N., and Casseus, R. G.: 1962, *J. Food Sci.* **20**, 560.
[84] Locker, R. H.: 1960, *Food Res.* **25**, 304.
[85] Herring, H. K., Cassens, R. G., and Briskey, E. J.: 1965, *J. Sci. Food Agr.* **16**, 379.
[86] Love, R. M.: 1958, *Nature* **182**, 108.
[87] Reay, G. A. and Kuchel, C. C.: 1936, *Rep. Food Invest. Bd. London*, p. 93.
[88] Hotani, S.: 1956, *J. Tokyo Univ. Fish.* **42**, 89.
[89] Love, R. M. and Elerian, M. K.: 1964, *J. Sci. Food Agr.* **15**, 65.

[90] Olley, J. and Duncan, W. R. H.: 1965, *J. Sci. Food Agr.* **16**, 99.
[91] Kishimoto, A. and Fujita, H.: 1956, *Bull. Jap. Soc. Sci. Fish.* **22**, 293.
[92] Stering, C.: 1955, *Food Res.* **20**, 474.
[93] Reeve, R. M. and Notter, G. K.: 1954, *Food Technol.* **13**, 574.
[94] Unrau, A. M. and Nylund, R. E.: 1957, *Am. Potato J.* **34**, 245.
[95] Kuhn, G., Desrosier, N. W., and Ammerman, G.: 1959, *Food Technol.* **13**, 183.
[96] Beachell, M. H.: 1959, *The Chemistry and Technology of Cereals as Food and Feed*, Avi Publishing Co.
[97] Rao, B. S., Vasuveda Murthy, A. R., and Subramanya, R. S.: 1952, *Proc. Indian Acad. Sci.* **368**, 70.
[98] Deshikacher, A. S. R. and Subrahmanyan, V.: 1961, *Cereal Chem.* **38**, 356.
[99] Chikubu, S., Horiuchi, H., and Tani, T.: 1958, *J. Jap. Soc. Agri. Chem.* (in Japanese) **32**, 268.
[100] Chikubu, S. and Horiuchi, H.: 1964, *Foodstuffs* issued by Institute for Food Research (in Japanese) **7**, 14.
[101] Shimizu, T., Fukawa, H., and Ichiba, A.: 1958, *Cereal Chem.* **35**, 34.
[102] Harris, R. H. and Sibbitt, L. D.: 1958, *Food Technol.* **12**, 91.
[103] Steiner, E. H. and Mill, C. C.: 1959, *Rheology of Disperse System*, Pergamon Press, New York, p. 171.
[104] Sterling, C. and Wuhrmann, J. J.: 1960, *Food Res.* **25**, 460.
[105] Davis, J. G.: 1937, *J. Dairy Res.* **8**, 245.
[106] Hunziker, O. F., Mills, H. C., and Spitzer, G.: 1912, *Ind. Exp. Stat. Bull.* **159**, 283.
[107] Coulter, S. T. and Hill, D. J.: 1934, *J. Dairy Sci.* **17**, 543.
[108] Sterling, C. and Wuhrman, J. J.: 1960, *Food Res.* **25**, 460.
[109] Feuge, R. O. and Guice, W. A.: 1959, *J. Am. Oil Chem. Soc.* **36**, 531; Lovegren, A. V., Guice, W. A., and Funge, R. C.: 1958, *J. Am. Oil Chem. Soc.* **35**, 327.
[110] Huebner, V. R. and Thomson, L. L.: 1957, *J. Dairy Sci.* **40**, 834, 839.
[111] Mohr, W. and Drachenfels, H. J.: 1955, *Fette Seifen Anstrichmittel* **11**, 925.
[112] de Man, J. M. and Wood, F. W.: 1959, *J. Dairy Res.* **26**, 17.
[113] Van den Tempel, M.: 1961, *J. Colloid Sci.* **16**, 284.
[114] Mulder, H.: 1940, *Versl. Landbk. Onderz.* **46**, 21.
[115] Burger, J. M. and Scott Blair, G. W.: 1949, *Rep. Principle of Rheol. Nomencl.*, North-Holland Publishing Co.
[116] Avrami, M.: *J. Chem. Phys.* **7** (1939), 1103; **8** (1940), 212; **9** (1941), 177.
[117] Finke, A. and Heinz, W.: 1956, *Fette Seifen Anstrichmittel* **58**, 905.
[118] Harbard, E. H.: 1956, *Chem. Ind.*, p. 491.
[119] Loska, S. J. Jr. and Jaska, E. J.: 1957, *J. Am. Oil Chem. Soc.* **34**, 495.
[120] Dolby, M.: 1941, *J. Dairy Res.* **12**, 344.
[121] Kapsalis, J. G., Betsher, J. J., Kristofersen, T., and Gould, I. A.: 1960, *J. Dairy Sci.* **43**, 1560.
[122] Nickerson, T. A. and Pangborn, R. M.: 1961, *Food Technol.* **15**, 105.
[123] Nickerson, T. A. and Pangborn, R. M.: 1961, *Food Technol.* **15**, 105.
[124] Leighton, A., Leviton, A., and Williams, O. E.: 1934, *J. Dairy Sci.* **17**, 639.
[125] Sone, T. and Taneya, S.: 1966, *Proc. 17th Intern. Dairy Congr.* E. 3.
[126] Shaw, D.: 1963, in P. Sherman (ed.), *Rheology of Emulsions*, Pergamon Press, London.
[127] Stistrup, K. and Andreasen, J.: 1962, *Proc. 16th Intern. Dairy Congr.* C. 29.
[128] Taneya, S.: 1963, *Physics of Powdered Milk in Dairying Technology* Ser. 2 (in Japanese), Asakura Publishing Co.
[129] Taneya, S. and Sone, T.: 1962, *Appl. Phys.* (in Japanese) **31**, 286.
[130] Taneya, S. and Sone, T.: 1962, *Appl. Phys.* (in Japanese) **31**, 483.
[131] Taneya, S. and Sone, T.: 1962, *Appl. Phys.* (in Japanese) **31**, 465.
[132] Matheson, G. L., Herbest, W. A., and Holt, P. H. Jr.: 1949, *I.E.C.* **41**, 1099.
[133] Benarie, M. M.: 1961, *British J. Appl. Phys.* **12-9**, 514.
[134] Ono, E. and Taneya, S.: 1963, *Powder Powdered Metallurgy* (in Japanese) **10**, 19.
[135] Taneya, S. and Sone, T.: 1962, *J. Soc. Material Sci. Japan* (in Japanese), **12**, 300.
[136] Athy, L. F.: 1930, *Bull. Am. Soc. Petrol. Geol.* **14**, 403.
[137] Walker, E. E.: 1923, *Trans. Faraday Soc.* **19**, 404.
[138] Kawakita, K.: 1956, *Ann. Rept. Takamine Res. Lab.* (in Japanese) **8**, 83.
[139] Taneya, S. and Takada, K.: 1963, *Powder Powdered Metallurgy* (in Japanese) **10**, 15.
[140] Chuchlowa, J.: 1958, *Dairy Sci. Abstr.* **20**, (7) 563.
[141] Tanaka, T. and Kawai, S.: 1956, *Chem. Eng.* (in Japanese) **20**, 144.

[142] Raush, J. M.: 1949, *Gravity Flow of Solids Beads in Vertical Towers*, Ph.D. Thesis, Prinston Univ.
[143] Van de Velde: 1961, *Chem. Eng. Sci.* **15**, 260.
[144] Flowler, R. T. and Glastonbury, J. R.: 1959 *Chem. Eng. Sci.* **10**, 10.
]145] Duck, W. N.: 1961, *Confectionary Production* **27**, 153.
[146] Tiemstra, P. J.: 1964, *Food Technol.* **18**, 921.

CHAPTER 4

ABNORMAL FLOW PROPERTIES OF FOODSTUFFS

4.1. Dependence of Food Consistency Upon Time

In general, foodstuffs cannot be stored for a long period of time so it is necessary to obtain information on the denaturation of materials during storage. The important thing is to endeavour to preserve foodstuffs without denaturation for as long as possible.

Many techniques for processing foods have been developed, e.g. desiccation, salting, sugaring, pasteurization, canning, etc. Canned foods contain a large quantity of water in which are protein molecules denatured by the thermal treatment. Moreover, the denaturation of protein continues during storage, and this alters the consistency of the system due to interaction between the denatured protein molecules and the other components of the food. The subject of this chapter will be deeply concerned with consistency changes of foodstuffs during storage.

4.1.1. INCREASE OF VISCOSITY OF CONDENSED MILK

As a generalisation it can be said that condensed milk is prepared by the concentration of milk with sugar. Bacteria hardly breed in condensed milk, because this product is eventually packed in a can, and also because the osmotic pressure in the system becomes very high with increasing concentration of sugar. Therefore, any increase in the viscosity of condensed milk may be mainly due to physico-chemical action. Hunziker [1] has reported that the hydration of protein affects the viscosity of condensed milk, and other investigators [2] have suggested that the major cause of viscosity change may be association between the casein micelles under the influence of calcium ions in the system. Using an electronmicroscope, Hostettler and Imhof [3] actually observed non-spherical coagulated protein masses.

Condensed milk is a very complicated suspension, in which colloidal protein, fine crystals of lactose, and fat globules are dispersed in a saturated solution of lactose and sucrose. The viscosity of condensed milk was studied a long time ago by Stebnitz and Sommer [4] using a falling sphere viscometer, and by Webb and Hufnagel [5] using a MacMichael viscometer, but the dependence of viscosity upon rate of shear was not fully investigated at that time. More recently, Samel and Muers [6] have tried to measure the viscosity of condensed milk at various rates of shear at 20 °C with a Ferranti portable viscometer. This viscometer is basically a Couette type viscometer except that the bob is immersed in the liquid sample in order to avoid local disturbance of the sample. These workers actually employed condensed skim milk as the sample, because it is thicker than condensed whole milk.

Flow curves of thickened condensed milk are compared with that of a freshly prepared sample in Figure 4.1. Although the fresh sample shows nearly Newtonian flow, thickened condensed milk has a yield value and exhibits plasticity. Figure 4.1 also illustrates up- and down-hysteresis loops in the flow curves, which have been obtained by different rotation times at a constant rate of shear. With increasing rotation time at a constant shear rate, the viscosity of thickened condensed milk decreases and approaches that of the freshly prepared sample.

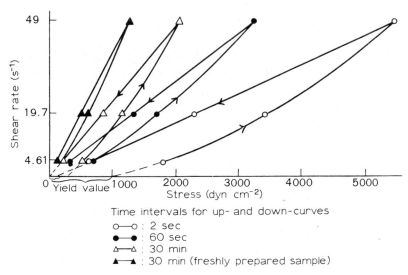

Fig. 4.1.   Dependence of rotation period of viscometer at a fixed speed upon flow curve of thickened condensed milk.

There are two possible factors for the thickening of condensed milk during storage; one of them is induced by the nature of colloidal protein, and the other is the presence of sugar crystals. The latter factor can be neglected, because the yield value of gelatin solution does not change in the presence of crystals of sucrose and lactose. Therefore, it appears that the colloidal protein dispersed in the system plays a big role in the viscosity change of condensed milk, i.e. thickening during storage and thinning under the influence of external disturbance. Whey protein is not important for the thickening of milk, because one cannot identify an increase of viscosity in the condensed whey system. This is confirmed by the fact that if casein protein is fractionated out from thickening condensed milk by centrifugation, the viscosity of the supernatant fluid has the same value as that of supernatant derived from freshly prepared condensed milk, as shown in Figure 4.2.

The rate of casein protein fractionation from thickened condensed milk using a centrifuge is faster than from a freshly prepared sample, and this could be due to an increase of geometrical symmetry in the casein micelles or an increase in the density of the micelles in the thickened condensed milk. The former speculation is incompatible with the thickening process, and the latter may cause a decrease in the

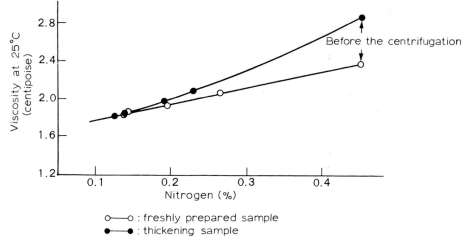

Fig. 4.2.   Effect of centrifugation of diluted condensed milk.

viscosity due to dehydration of the micelles. So a reasonable explanation would be an increase of micellar mass brought about by coagulation between the micelles, and thus forming a very complicated shape.

From the above discussion and the experimental results described, the thickening of condensed milk may be induced by two factors; first, an irreversible change in the molecular structure of protein, and second, the formation of a weak network structure between the protein micelles, which is easily broken by mechanical force. The first contributory factor may also promote the formation of a network structure in the system, so that a thixotropic gel gradually appears in condensed milk during storage.

The rate of thickening of condensed milk is affected by the temperatures of both condensation and storage. The rate of thickening increases with decreasing condensation temperatures ranging from 22 °C to 55 °C, but thickening in a low temperature range can be eliminated by preheating at 85 °C for 10 min. Therefore, if the milk is concentrated at a relatively high temperature, and it is then stored at a relatively low temperature, various insoluble salts, especially calcium phosphate, should precipitate from the system, so that the condensed milk will be stable during storage. It follows that preheating promotes the nucleus formation of salts, which are then separated from the condensed milk during condensation and storage.

Samel and Muers [7] have reported that the thickening of condensed milk is pronouncedly accelerated by the addition of salts. Samel and Muers [8] have also studied the anion and cation dependence of the thickening process, and they reported that polyvalent anions form a strong bridge together with calcium atoms in the calcium caseinate molecule. The bridging effect contributes to casein coagulation in the system. On the other hand, the thickening process is retarded in the presence of fluorides, because fluorides are monovalent anions and they inhibit bridge formation with the calcium atoms. The stabilizing effect of cations on the thickening of condensed milk appears to be due to interaction with polyvalent anions (phosphoric acid ion, citric acid ion, etc.) in the system.

Thus, one may conclude that the cause of the thickening of condensed milk can be separated into two individual parts, i.e. an irreversible thickening process due to denaturation of protein and the formation of a thixotropic gel structure due to the weak network structure of the protein micelles. Both phenomena are brought about by structural changes within the protein molecules.

4.1.2. EFFECT OF WORKING ON THE CONSISTENCY OF EDIBLE FATS

Van den Tempel [9] has studied the compression of cylindrically moulded plastic fat, which was prepared by mixing molten tri-stearin with paraffin oil. From the results obtained he deduced that the fat crystals form a scaffold structure involving both reversible weak bonds and irreversible strong bonds, and that irreversible softening of the system produced by isothermal working is due to rupture of irreversible strong bonds, while the thixotropic behaviour of the system is caused by the presence of reversible weak bonds between the fat crystals.

Margarine is usually prepared by rapidly cooling a water-in-edible oil emulsion, and plasticity appears in the system because of partial crystallization of fat component after the cooling. Fat crystal size in margarine is generally below one micron. Figure 4.3 shows an electron microphotograph of margarine. Although it is difficult to follow the change of fat crystal size, a large proportion of the crystals may withstand the pressure developed during working.

Haighton [10] has studied the change of hardness in margarine, butter and shortening after various degrees of working and setting in an isothermal condition using a cone penetrometer. The term yield value was employed to describe the hardness of the samples. Thus, the degree of softening due to the working W can be given by

$$W = \frac{C_u - C_w}{C_u} \times 100\,(\%), \tag{1}$$

where C_u and C_w are the yield value of the sample before working and after working, respectively. The relationship between the depth of penetration and the yield value of the sample is empirically expressed by the following relation;

$$C = KM/p^n, \tag{2}$$

where C is the yield value (g cm^{-2}), M is the penetrometer load (g), p is the distance of penetration (mm), n is the index taken as 1.6 for margarine, butter and shortening, and K is a constant which depends on the cone angle of penetrometer. We can derive the correlation between the distance of penetration and the degree of softening from Equation (1), as follows;

$$W = \left[1 - \frac{p_u^{1.6}}{p_w} \right] \times 100\,(\%), \tag{3}$$

where p_u and p_w are the distances of penetration in the sample before working and after working, respectively.

Fig. 4.3.   Electron micrograph of margarine.

In the manufacture of edible fats, some treatments such as rapid cooling, mixing, settling, etc. promote crystallization of the fat components in the system, and the fat crystals then a network structure in the system. It seems that Van der Waals-London forces play a big part in the formation of the network structure, so that the distance between the elementary crystals in the structure may be below 0.01 μ. However, the structure is easily broken by mechanical treatments such as mixing, working, etc., and it is reconstructed during setting. This type of structure is described as 'secondary binding'; 'primary binding' describes a structure, which is never reconstructed by setting after working. The presence of primary bonds may be responsible for brittleness in fat systems.

TABLE 4.1

Decrease of sample hardness with multiple working

Sample	Initial yield value (g cm^{-2})	Yield value after working (g cm^{-2})			Degree of softening		
		1st	2nd	3rd	1st	2nd	3rd
Margarine I, 15 °C	1090	195	165	140	82.1	84.8	87.1
Margarine II, 15 °C	760	245	230	–	67.8	69.8	–
Shortening, 20 °C	2300	1300	1190	910	43.4	48.2	60.3
Shortening, 20 °C	2250	1120	1060	970	50.3	52.9	56.9

Haighton [10] obtained data on the change of yield value and softening of edible plastic fats due to working, as shown in Table 4.1, in which one can see that the degree of softening depends upon the kind of edible fat, e.g. the degree of softening is 70% ~ 90% for margarine and about 60% for shortening. Haighton [10] has also observed that the degree of softening for butter is 50% ~ 60%. Therefore, it follows that the spreadability of margarine on bread should be better than that of butter, even though the hardness of margarine is identical with that of butter before working. The degree of softening for shortening makes them suitable for use in cake making.

The isothermal reconstruction of secondary bonds in margarine and butter proceeds non-linearly against the logarithm of working time, but the relationship is linear for shortening. This variation in the reconstruction of the different fats after working may be brought about by differences in the number ratio of primary bonds to secondary bonds. Sone [11] has considered the change in the hardness of butter due to working and setting from the viewpoint of crystallization of the fat components, and they have found by measuring the change in density and viscosity, that the habit of fat crystallization in butter is deeply concerned with the statistical theory of crystal growth.

In butter the spherical fat crystals are suspended in a continuous phase of liquid oil, but margarine is composed of non-spherical fat crystals in a liquid oil. The dependence of the dispersion state of fat crystals, including crystal shape, upon the consistency of plastic fat systems should be investigated in future work together with studies on the presence of fat crystals in the submicron range in butter.

4.2. Surface Consistency of Foods

The surfaces of foods are exposed to air, or are in contact with the package, so that denaturation can often be seen more on the surface than on the inside. Both physical and chemical denaturation occur on the surfaces of foods. The surfaces also generally show the phenomenon of adhesion.

4.2.1. STICKINESS OF CANDY

When candy is stored in a moist atmosphere, either stickiness develops on the surface, or a crystallized turbid layer gradually spreads into the inside from the surface. The latter phenomenon can be divided into three stages, as follows:

(1) Crystallization commences following deliquescence on the surface with a relatively low moisture concentration.

(2) An equilibrium state develops in the crystalline phase with the saturated deliquescent phase.

(3) The deliquescent phase gradually inverts to an unsaturated sugar solution on the surface of candy with a relatively high moisture content.

The surface state in each stage described above is different to the others, so that the stickiness of candy is also different. Heiss [12] designed an apparatus for measuring stickiness (see Chapter 2), and he tried to determine the surface consistency of candy in a range of relative humidities from 30% to 70% at a temperature of 20 °C using a fixed moisture content and storage period. The stickiness on the surface layer of candy is much influenced by the sugar concentration of the deliquescent phase, so that the stickiness varies according to the surface condition of candy.

In the intermediate region of the process, a layer of fine particles gradually appears, and then crystallization commences, while stickiness decreases. The rate of appearance of fine particles increases with increasing relative humidity, and maximum stickiness appears twice, as shown in Figure 4.4. The first maximal stickiness can be seen just before the commencement of crystallization, then the second peak appears in a region where the crystal lattice becomes weak. However, it is difficult to obtain the second peak in a condition of relatively low humidity.

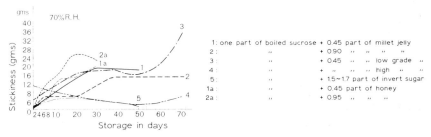

Fig. 4.4.   Rate of appearance of particle layer in candy.

It is easy to obtain fine particles in a relative humidity of 35%, when candy is prepared from invert sugar. The stickiness of candy is much affected by the ash content and impurities in the sample, i.e. the stickiness increases with increasing content of both ash and invert sugar.

In order to maintain a good quality of candy, it is necessary to delay the attainment of the first maximal value of stickiness by controlling the appearance of the fine particles and crystals. This can be obtained by using a blend of highly viscous material or by protecting the sample from humidity by packaging materials.

4.2.2. STICKINESS OF BUTTER

Butter is constructed from a scaffolding structure of fat crystals, and it is a plastic body. The process of working destroys the scaffolding structure, and butter then becomes a pasty body.

Jansen [13] has tried to measure the combined forces of adhesion and cohesion (in gramme units) by pulling a metal cylinder away from the surface of butter. He also studied the correlation between stickiness and the extrusion parameters of butter such as extrusion force, friction coefficient, degree of slip, etc. These parameters were obtained using an extruder. Table 4.2 summarizes the correlation coefficients

TABLE 4.2
Correlations between stickiness and various extrusion parameters of butter

	Extrusion force	Friction coefficient	Degree of slip	Syneresis oil
	Correlation coefficient at 10°C			
Stickiness	−0.76	−0.59	+0.58	+0.36
Extrusion force	−	+0.35	−0.56	−0.26
Friction coefficient	−	−	−0.88	−0.44
Degree of slip	−	−	−	+0.63
	Correlation coefficient at 15°C			
Stickiness	+0.07	+0.33	−0.25	−0.11
Extrusion force	−	+0.79	−0.82	−0.72
Friction coefficient	−	−	−0.89	−0.82
Degree of slip	−	−	−	+0.98
	Correlation coefficient at 20°C			
Stickiness	+0.86	+0.65	−0.79	−0.71
Extrusion force	−	+0.94	−0.79	−0.72
Friction coefficient	−	−	−0.89	−0.69
Degree of slip	−	−	−	+0.84

obtained, and one can see that the correlation coefficients are much influenced by the temperature at which measurements were made. The stickiness of butter correlates little with factors affecting the extrusion of butter at 15°C, but a correlation is found between the stickiness and the extrusion pressure at temperature of 10°C and 20°C, respectively.

From the experimental results obtained by Jansen [13], it is possible to conclude that the first working increases the stickiness of butter, but subsequent workings do not influence stickiness. Recently, Thomasos and Wood [14] have compared the stickiness of ordinary butter with that of so-called 'golden flow butter' prepared by means of a continuous machine. They have also investigated the influence of re-working or air content upon the stickiness of butter.

Figure 4.5 shows the effect of reworking on hardness, stickiness and amount of syneresis oil of ordinary and 'golden flow' butters. In the case of 'golden flow' butter both hardness and the amount of syneresis oil gradually decrease with working, but the hardness of ordinary butter is not so much affected by working. The stickiness of ordinary butter increases initially and then decreases with repeated working, but 'golden flow' butter only increases in stickiness with repeated working.

It should be mentioned that the stickiness of butter is inversely proportional to the amount of syneresis oil, as shown in Figure 4.5, and it decreases when air is in-

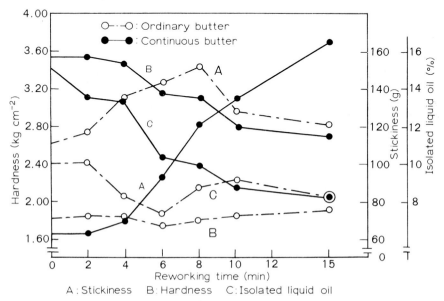

Fig. 4.5. Dependence of reworking upon stickiness, hardness and amount of isolated liquid oil of ordinary butter and continuous butter.

corporated. The aeration does not influence the cohesion and adhesion of butter, but it reduces the hardness. However, if a large number of air bubbles are dispersed in butter, the apparent cohesion may decrease, so that a large part of the butter surface may stick to the wrapper.

References

[1] Hunziker, O. F.: 1949, *Condensed Milk and Milk Powder*, 7th ed, Illinois.
[2] Samel, R. and Muers, M. M.: 1962, *J. Dairy Res.* **29**, 269.
[3] Hostettler, H. and Imhof, K.: 1953, *Proc. XIIIth Intern. Dairy Congress* **2**, 423.
[4] Stebnitz, V. C. and Sommer, H. H.: 1935, *J. Dairy Sci.* **18**, 757.
[5] Webb, B. H. and Hufnagel, C. F.: 1948, *J. Dairy Sci.* **31**, 21.
[6] Samel, R. and Muers, M. M.: 1962, *J. Dairy Res.* **29**, 249.
[7] Samel, R. and Muers, M. M.: 1962, *J. Dairy Res.* **29**, 259.
[8] Samel, R. and Muers, M. M.: 1962, *J. Dairy Res.* **29**, 269.
[9] Van den Tempel, M.: 1958, *Rheol. Acta* **1**, 115.

[10] Haighton, A. J.: 1959, *J. Am. Oil Chem. Soc.* **36**, 345.
[11] Sone, T.: 1961, *J. Phys. Soc. Japan* **16**, 961.
[12] Heiss, R.: 1959, *Food Technol.* **13**, 433.
[13] Jansen, K.: 1961, *J. Dairy Res.* **28**, 15.
[14] Thomasos, F. I. and Wood, F. W.: 1964, *J. Dairy Res.* **31**, 137.

CHAPTER 5

SENSORY ASSESSMENT OF FIRMNESS

5.1. Psychorheology

The mechanical properties of foodstuffs which are mainly dispersed systems consisting of many components are usually described by the term consistency or firmness. It is well known that Japanese people use some unique terms for describing the mechanical properties of foodstuffs such as 'koshi', 'ashi', 'hagire', etc.; these terms are of course familiar to Japanese people, and they correspond with the sensory responses to the stimuli. However, the meaning of each sensory response differs from the others, because the stimuli take various forms such as finger, tongue, mastication, etc. Also the degree of sensory response differs with the individual.

On the other hand, rheology is characterized by the study of objective correlations between stress, strain, and time for materials using the physical properties of viscosity and elasticity. Dr Scott Blair*, of the National Institute for Research in Dairying, Reading, England, developed the field of 'psychorheology' for obtaining information on the correlation between the psychological and subjective aspects of the mechanical properties of materials and the physical and objective aspects of rheology. Dr Scott Blair initiated a unique study on the sensory assessment made by craftsmen in cheese making or baking.

5.2. Differential Threshold Between Elasticity and Viscosity

Foodstuffs are generally complex materials which exhibit both viscosity and elasticity in their mechanical behaviour, so that when the firmness of foodstuffs is assessed subjectively it may not be clear whether one is judging viscosity or elasticity. As viscosity and elasticity are the objective quantities for characterizing the different mechanical dimensions of materials, it is possible that viscosity gives a different stimulus to the sensory perception from elasticity.

As far as the sensory response to the psychological stimulus is concerned, we have Weber-Fechner's law, which states that the degree of sensory response is proportional to the logarithmic strength of the stimulus. We also have the terms 'differential threshold' and 'stimulus threshold'; the former means a minimum boundary for identifying subjectively the difference between viscosity and elasticity, and the latter corresponds to the critical value of judgeable stimulus.

---

* The translator notes that Dr Scott Blair retired from N.I.R.D. in 1967.

Scott Blair and Coppen [1] obtained information about the subjectively judgeable difference in viscosity and in elasticity in the light of Weber-Fechner's law. They prepared five samples from a mixture of Newtonian fluids such as California bitumen and oil; the viscosity of these samples ranged from 10^6 poise to 10^7 poise. Then, the panels in the above experiment were asked to provide a ranking order on the firmness of the samples by squeezing them in water at $18\,°C$ as viscosity is much affected by temperature. Thirty panels were divided into six groups, and each panel attempted the above test five times with each sample (twenty-five times for all the samples), so that seven-hundred and fifty answers were collected from all the panels. These answers were classified into three categories according to the age, standard of education, and occupation of the panelists, and the sensory sensitivity of each category was determined as the percentage of correct answers by means of the so-called 'chi-square test'. The results obtained showed that a group of chemical analysts were generally very sensitive in their judgements, but that both groups with a high standard of education and of advanced age did not give very sensitive results. A significant point was that the responses obtained from craftsmen in the cheese industry, who assessed the firmness of cheese with their fingers every day, were not always correct in this test. In general, eighty percent of the panelists gave correct answers, when the percentage difference in viscosity between samples was 30%.

Scott Blair and Coppen also prepared a series of vulcanized rubber samples so as to determine the judgeable boundary of elasticity. The cylindrical samples, whose height and diameter were 2.5 cm and 2 cm, respectively, had a series of bulk moduli in a range of $1 \sim 2 \times 10^7$ dyn cm^{-2}. The test was made with ten panels by compressing the samples between the thumb and forefinger. Although the individual ability of perception was obviously different with each panelist, it was possible to classify the ability of perception according to the same category of panelists as for the viscosity test. The responses from craftsmen in the cheese industry were also inferior to those of the other panelists. It generally followed that eighty percent of correct answers were obtained when the percentage difference in elasticity between samples was ten percent.

Figure 5.1 shows the threshold curves for elastic moduli and viscosities, i.e. plots of percentage correct answers against percentage difference in viscosity or modulus for paired samples, from which it is obvious that sensory perception of elasticity is more sensitive than for viscosity. Scott Blair also tried to use pairs of samples such as a steel spring and a rubber cylinder for the sensory evaluation of elasticity, and he deduced from the results that subjective evaluation of elasticity by the squeezing test can be carried out more sensitively with a rubber cylinder than with a metal spring. Accordingly, it can be stated that sensory perception on the skin plays a big part in the overall sensory assessment.

Let us suppose that the sensory assessment made with a fixed squeezing force depends on the degree of deformation of the sample within a fixed period of time, then if the firmness of a viscous fluid is compared with that of an elastomer by squeezing for just one second, one may possibly judge the two samples to have equal

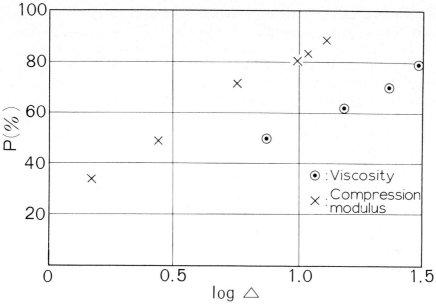

Fig. 5.1.   Threshold curves for elastic moduli and viscosities (P is percentage correct, and $\varDelta$ is percentage difference in viscosity or modulus of the pair).

firmness in the case where the viscosity coefficient and elastic modules have identical values. The physical dimension of viscosity is $ML^{-1}T^{-1}$, and that of elastic modulus is $ML^{-1}T^{-2}$, so that the comparison of viscosity with elasticity involves a comparison over a time scale. As a matter of fact, the squeezing force applied to the sample is not a significant factor in the sensory assessment, but the time for which the squeezing force is applied plays a big part in the sensory evaluation of firmness for a pure viscous fluid or a pure elastomer.

The firmness of a viscous material is subjectively judged by the rate of flowing deformation, so that one makes an assessment of viscosity dynamically. On the other hand, the assessment of elastomer firmness is made statically.

5.3. Perception of Firmness of Complex Materials [2, 3]

The term 'complex material' as used in this section corresponds to a viscoelastic body, which means that the material possesses properties intermediate between viscosity and elasticity.

In order to prepare such complex materials, eight mixtures of gum, vaseline and clay were moulded into cylindrical forms whose weight, diameter and height were 222 g, 4.5 cm and 8 cm, respectively. Then, the ranking order of firmness of these samples was made with nineteen panelists, who were required to squeeze each sample twenty-five times. The twenty-five responses per sample were subdivided into the initial thirteen and the final twelve, and the variance in the ranking order of the respective subdivisions was calculated as σ_1 and σ_2, and the ratio $\sigma_1/\sigma_2 = \varLambda$ was defined as the proficiency factor. Craftsmen in the cheese industry gave very irregular answers again in this test, i.e. significant fluctuations could be seen in the variance σ_1 and σ_2, and Λ was also inferior. It appeared from the above results that the craftsmen tired to such an extent in tests with unfamiliar samples that their judgement were irregular, although they showed proficiency in the assessment of cheese firmness.

Let us suppose that the subjectively determined ranking order is a, b, c, d, e, f, g, and h according to a sequence of firmness from hard to soft, and if the degrees of deformation under compression with a fixed load within a fixed period of time for the hardest sample a and the softest sample h are A and H, respectively, it is possible to follow the differences in firmness between samples from the results of the above ranking order. For example, the objective difference of firmness between samples d and e can be estimated by the following relation,

$$D - E = (A - H)(d - e)/(a - h). \tag{1}$$

Figure 5.2 shows the plots of percentage correct responses (P) against objectively determined difference of firmness on a logarithmic scale $(\log \Delta)$. The threshold curve for firmness of a complex material tends to be identical with that of a pure viscous fluid such as bitumen. Although cheese falls within the category of complex materials, one may scarcely expect to have good answers on the firmness of other complex materials form the craftsmen in the cheese industry.

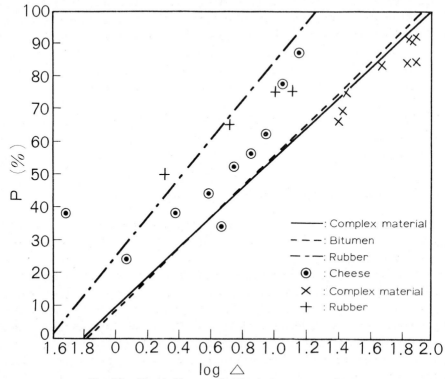

Fig. 5.2.   Threshold curves for simultaneous comparisons.

5.4. Mechanical Properties and Firmness in Subjective Perception [4, 5]

The force applied to a sample is represented quantitatively by the term stress per unit area of the sample. The stress can be divided into two components one of which is a vertical component and the other is tangential to the surface of the material. Deformation is represented quantitatively by the term strain, which is expressed by the normal strain, shear strain, etc.

The most basic equations of rheology for an ideal material are Hookean elasticity and Newtonian viscosity. The relationship between the stress S and the strain σ for a Hookean system is given by

$$S = G\sigma, \tag{2}$$

where G is the elastic modulus. The viscosity of a Newtonian liquid is defined by the equation

$$S = \eta\,(d\sigma/dt), \tag{3}$$

where η is the viscosity coefficient, and the term $d\sigma/dt$ is the rate of strain.

A number of foodstuffs generally show viscoelastic behaviour, i.e. viscous flow, retarded elasticity and instantaneous elasticity within a given period of time under a fixed stress. Although such behaviour can be interpreted using the combined mechanical models of Hookean elasticity and Newtonian viscosity, this approach is only useful for interpreting the mechanical response of idealized materials.

Scott Blair has proposed a generalized equation for interpreting the mechanical properties of real complex materials, as follows:

$$\psi = S^\beta \sigma^{-1} t^k, \tag{4}$$

where ψ varies with indices β and k. The simplest cases are the following:

$$\psi = G, \quad \text{when } \beta = 1 \text{ and } k = 0$$
$$\psi = \eta, \quad \text{when } \beta = 1 \text{ and } k = 1.$$

If the stress is kept constant, the relationship between log strain and log time should be linear, as shown in Figure 5.3, so that

$$\log \psi = \log S - \log \sigma + k \log t. \tag{5}$$

The above equation had previously been put forward by Nutting, and Scott Blair and Coppen examined this equation with regard to its suitability for describing the firmness of complex materials from the viewpoint of sensory perception.

First, they provided a pure viscous fluid prepared from California bitumen and a series of pure elastic samples prepared from rubber, respectively. The viscosity of California bitumen was η, and the elastic moduli of the rubbers were n_1, n_2, n_3 and n_4. The panelists used in this test had to squeeze each sample for different periods of time such as a half second, one second, two seconds and four seconds, and they were then requested, after squeezing each sample with the left hand and then with the right

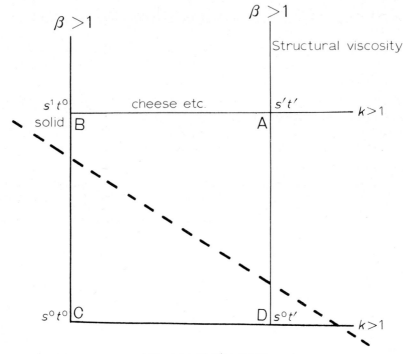

Fig. 5.3. Nutting square.

hand, to state which appeared softer the bitumen or the rubbers. A linear relationship
was obtained between the percentage of panelists, who answered that the rubbers
were softer than the bitumen, and the $\log(nt_c)$, where t_c is the time for which a
sample was squeezed, as shown in Figure 5.4. Such a tendency may suggest that the
squeezing time t_c plays a big part in the sensory evaluation of elastic materials. It
may also be assumed that if the stimulus for viscosity is identical with that for
elasticity, then an intersecting point should be obtained at fifty percent panelists
against the value nt_c. This represents a numerically equivalence of elasticity with
viscosity. However, the intersecting point appears at a point where nt_c deviates
negatively from the above assumption, as shown in Figure 5.4. Therefore, Equation
(4) should be rewritten as

$$\psi_c = S\sigma^{-1}(dt)^k, \tag{6}$$

where d corresponds to the psychological time.

In a further study Scott Blair and Coppen prepared a series of samples by mixing
together clay and vaseline with crude rubber in various ratios, and they selected
a standard sample from the above preparations. It proved possible to represent the
mechanical properties of all the samples by Nutting's equation, in which the term ψ
differed for each sample, but the term k remained the same for all samples. Sensory
evaluations were made with ten panelists, who were requested to answer the
question, "Which sample is softer, the standard or one of the others?". Each test

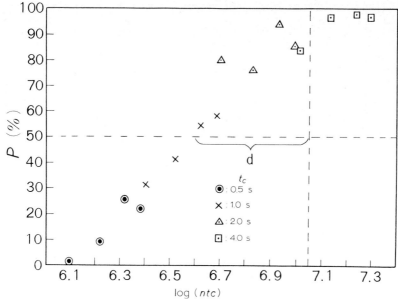

Fig. 5.4.   Results of simultaneous comparisons with two compressions.

was repeated three times, thus giving a total of thirty-two tests. The results obtained are summarized in Figure 5.5, where the percentage panelists P_u who answeres that the standard sample was softer than the other samples, is plotted against $\log(\psi_m t_c^{Ak})$. This plot appears to be linear, so that the strain σ_u in the standard sample after t_c second squeezing can be written as

$$\sigma_u = t_c^{k_u}\psi_u^{-1}S, \qquad (7)$$

where k_u and ψ_u correspond to the terms k and ψ for the standard sample in Nutting's equation. If one of the other samples shows a strain identical with that in the standard sample after t_c seconds squeezing, and if this strain σ'_m can be written as

$$\sigma'_m = t_c^{k'_m}\psi'^{-1}_m S \qquad (8)$$

the following relation should be derived,

$$t_c^{(k_u - k'_m)} = t_c^{Ak} = \psi_u/\psi'_m \qquad (9)$$

because $\sigma_u = \sigma'_m$. Equation (9) means that when the deformation of the sample after one second squeezing is the same as that of the standard, and if the two samples are believed to have identical firmness, then t_c^{Ak} should be inversely proportional to ψ'_m.

In Figure 5.5, the vertically dotted line corresponds to the firmness of the standard sample, and the horizontally dotted line is consistent with 50% of PP_u. If the level of sensory perception for the sample is the same as for the standard, the plot of P_u against $\log \psi_m t_c^{Ak}$ should pass across the point, where both dotted lines cross each

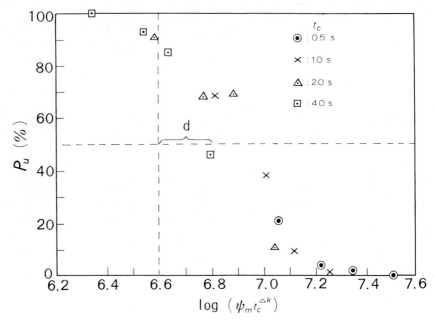

Fig. 5.5.   Percent standard softer answers in simultaneous comparisons.

other. However, the real crossed point deviates from the above by a value d, so that the psychological time can be calculated, as follows:

$$\log \alpha = \frac{\log(\psi_m t_c^{\Delta k}) - \log \psi_u}{\Delta k} = \frac{d}{\Delta k} \tag{10}$$

$$d = \Delta k \log \alpha \tag{11}$$

Figure 5.6 shows the experimentally obtained correlation between Δk and d, and one finds that d has its largest value in the case where $\Delta k = 1$, and that d increases with decreasing Δk for small values of Δk. An increase of Δk means that the mechanical properties of the comparative samples become different from one another. Therefore, the sensory perception of the firmness of materials is much affected by the presence of the term α, and the degree of deformation after t seconds squeezing is very important to the sensory assessment of firmness. When the firmness of rubber is compared with that of bitumen, it is possible to predict that the former will be assessed as harder than the latter; the deformation of rubber is instantaneously kept constant after applying the stress, but in the case of bitumen the deformation is continuously changing even after t_c second squeezing. If the degree of deformation of the standard is equivalent to that of the sample which is compared with it, and equation can be derived, as follows:

$$\psi_u = \frac{k_u}{k_m} \psi_m \cdot t_c^{\Delta k} \tag{12}$$

In other words, d is equivalent to $\log(k_m/k_u)$ for obtaining Equation (12). When the

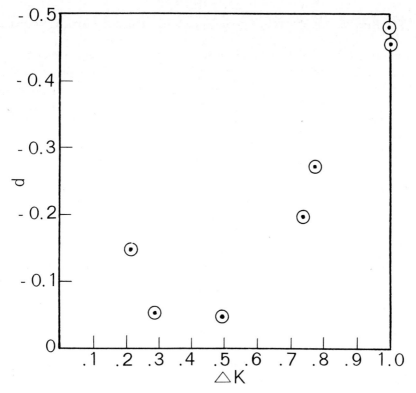

Fig. 5.6.   Correlation between d and Δk.

degree of deformation after t_c seconds is associated with the mean rate of deformation, it is possible to derive the following relation,

$$d\sigma/dt = k\,(\sigma/t) \tag{13}$$

Differentiation of the above equation gives the following relation:

$$\frac{d^2\sigma}{dt^2} = (k-1)\,\frac{1}{t}\,\frac{d\sigma}{dt} = k\,(k-1)\,\sigma/t^2 \tag{14}$$

$$\frac{d^n\sigma}{dt^n} = (k=1)\,(k-2)\ldots(k-n+1)\,t^{(k-n)}\,\psi^{-1}\,S$$

where $k\,(k-1)\,(k-2)\ldots$ is independent of the term $\psi_m t_c^{\Delta k}$, so that any order of differentiation maintains a linear relationship between $\psi_m t_c^{\Delta k}$ and P_u.

Physical techniques for measuring the firmness of materials are based on the deformation of the sample after unit time with a minimum value of usually one second. On the other hand, sensory perception is not affected by the degree of deformation after a period of time t_c, but it may be based on the relative rate of deformation during the time period t_c. Harper [8] has used this approach in cheese making, and he proved the above concept using the probit analysis.

5.5. Relationship Between Psychological Stimulus and Physical Stimulus

Schwartz and Foster [6] have published their ideas on the correlation between consumer's sensory perception of foodstuffs and the physical stimuli, as shown in Table 5.1. It is obvious from Table 5.1 that the comprehensive assessment of foods is

TABLE 5.1

Correlations between physical properties and sensory perceptions appeared in consumer

Physical stimulus	Sensory perception	Sensory response in consumer
Energies of reflection and radiation	Sense of sight	Colour, tissue, shape and size
Temperature	Sense of heat	Hot – cold
Viscosity	Sense of strength and pressure	Thick – thin
Density	Sense of strength and pressure	Heavy – light
Shear strength	Sense of strength and pressure	Hard, tough and soft
Moisture	Sense of pressure and coolness	Dry – wet
Vapour pressure	Sense of smell	Good or poor smell
Solubility	Sense of taste	Good or poor taste
Surface state	Sense of pressure and ache	Rough and prickly
pH	Sense of taste and ache	Sharp and sour

made not only with the sense of touch in the mouth or by hand but also with the sense of sight, temperature, smell, taste, etc. In view of the masticatory process in the mouth the texture profile may be divided into two categories; one, an initial perception of mastication such as geometrical assessment, and two, a mechanical assessment (Table 5.2). It has been further suggested that the perception of mastication can even be separated into three processes such as initial perception, intermediate mastication, and residual masticatory impression, respectively.

TABLE 5.2

Texture profile of foods in mastication

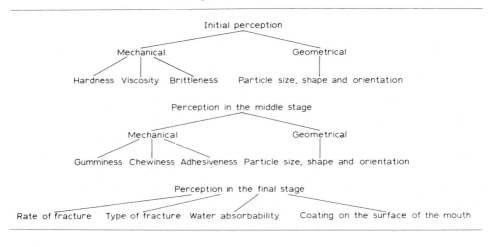

Szczesniak [7] has obtained good correlations between panel ratings of hardness, brittleness, chewiness, gumminess and adhesiveness, and the relevant measurements with a texturometer for foodstuffs, as shown in Figure 5.7. The texturometer, which has already been explained in Chapter 2 of this book, seems to be useful for obtaining good correlation between the relevant measurements and the subjective ratings; chewiness and adhesiveness show a linear relation with the objective measurement, and gumminess corresponds to the log texturometer readings. In addition to the above, the sensory term cohesiveness should be investigated so as to obtain experimental information on its correlation with the physical term elasticity, but it is difficult to provide standard samples for carrying out such experiment. Although it is possible to determine the texture profile associated with the initial perception of mastication using the second order variables of mechanics, a standard cannot be

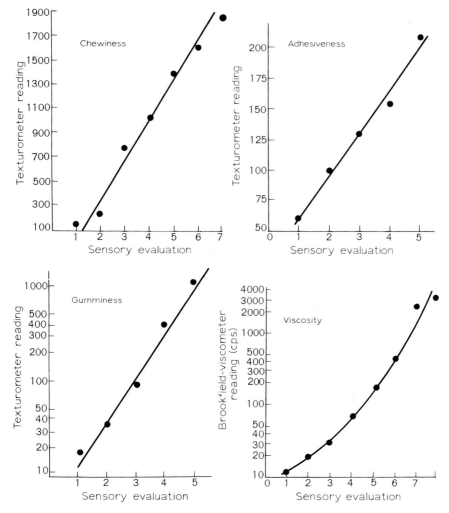

Fig. 5.7.   Correlations between relevant measurements and panel ratings.

provided for the process of perception in mastication because the material is cut and broken randomly by the teeth in this process.

As has been described above, Scott Blair *et al.* [4,5] defined a non-Newtonian time-scale for assessing the firmness of materials, while Szczesniak [7] provided some standards of texture and consistency, and they obtained correlations between the panel ratings of the various parameters and the relevant instrumental measurements. However, to establish correlations between psychological perception and physical measurements more circumstantial investigations are required. One should remember that lower reproducibility can be expected when obtaining physical informations on the firmness of complex materials such as foodstuffs. If the reproducibility of measurement is favourable, it should be possible to obtain useful information on the correlation between subjective assessment and the objective judgement. As a matter of fact, the consumer's comments on the quality of foods are generally very subjective, so that sensory assessment plays a big part in the development of food products. Food manufacturers should be required to analyse the details of sensory perception tests. Statistical studies have been made by Harper [8] in order to analyse the process of cheese making. This approach could usefully be extended to other foodstuffs.

The translator notes that, recently, Sherman [9] examined the Szczesniak texture profile, and suggested various modifications so as to place it on a more basic rheological foundation. He also indicated how it may be possible to predict panel response to sensory evaluation of texture from a limited number of physical measurements.

References

[1] Scott Blair, G. W. and Coppen, F. M. V.: 1939, *Proc. Royal Soc. B* **128**, 109.
[2] Scott Blair, G. W. and Coppen, F. M. V.: 1940, *Brit. J. Psychol.* **31**, 61.
[3] Coppen, F. M. V.: 1942, *Brit. J. Psychol.* **32**, 238.
[4] Scott Blair, G. W. and Coppen, F. M. V.: 1942, *Amer. J. Psychol.* **56**, 215.
[5] Scott Blair, G. W. and Coppen, F. M. V.: 1943, *Amer. J. Psychol.* **57**, 234.
[6] Schwartz, N. and Foster, D.: 1955, *Food Res.* **20**, 539.
[7] Szczesniak, A. S.: 1963, *J. Food Sci.* **28**, 385, 390 and 410.
[8] Harper, R.: 1952, *Psychological and Psycho-physical Studies of Craftsmanship in Dairying*, Cambridge Univ. Press. New York.
[9] Sherman, P.: 1969, *J. Food Sci.* **34**, 458.

CHAPTER 6

APPLICATION OF CONSISTENCY IN FOOD TECHNOLOGY

6.1. Consistency and Thermal Conductivity

During the processing of purees prepared from fruits and vegetables, the flow properties of the systems become increasingly non-Newtonian with changing temperature and concentration. Such problems play a very big part in the design of processing plants.

Harper [1] has studied the correlation between various viscosity parameters and thermal-conductivity in a rotary-fan type thin-layer evaporator using purees prepared from apricot, peach and pear. The shearing rate of non-Newtonian fluids against shear stress can generally be represented by

$$\tau = K\dot{\gamma}^n, \tag{1}$$

where τ is the shear stress, $\dot{\gamma}$ is the rate of shear, and K and n are constants, respectively. Apparent viscosity is generally expressed by $\eta_a = \tau/\dot{\gamma}$, and, therefore, Equation (1) can be rewritten as

$$\eta_a = \tau/\dot{\gamma} = K\dot{\gamma}^{n-1}. \tag{2}$$

The constants K and n can be measured experimentally by means of a coaxial cylinder viscometer, i.e. if $\log \tau$ is plotted against $\log \dot{\gamma}$ graphically, one may calculate the constant n from the gradient of the plot, and one may also find the constant K from the value of n at $\dot{\gamma}=1.0$. When $n=1.0$ and $K=\eta_a$, the fluid is Newtonian. Pigford [2] has proposed an equation for the thermal conductivity, as follows:

$$\frac{hD}{k} = 1.75 \left(\frac{wC_p}{kL}\right)^{1/3} \left(\sqrt{\frac{33n + 1}{4n}}\right), \tag{3}$$

where h is the heat transfer coefficient, D is the pipe diameter, k is the thermal conductivity, w is the flux of the system, C_p is the specific heat, L is the length of the pipe, and n is the power constant. Equation (3) fits the experimental results when the power constant n in Equation (1) has values from 0.4 to 1.0.

The votator heat exchanger is an apparatus for preparing edible fats such as margarine, shortening, etc., and a major use of this apparatus is for mixing molten fats with aqueous salt solution, and for cooling fats until they solidify. A votator heat exchanger consists of double pipe cylinders. The jacket of the outside cylinder is filled with coolant, while the inner cylinder can be rotated rapidly. The solidified fat system on the inside surface of the outer pipe cylinder is scraped off by means of symmetric blades, which are fixed to the outside surface of the inner cylinder. Although the viscosity of the fat should increase, and its non-Newtonian behaviour may become pronounced, as the degree of solidification increases, high speed rotation of the inner cylinder imposes such high shear on the fat system that the system may exhibit shear thinning. Skelland [3] has proposed an equation for representing the heat transfer coefficient of the system in a votator heat exchanger, as follows;

$$h_p = 4.9 \frac{k}{D_t} \left(\frac{D_t \bar{V} \varrho}{\eta_a}\right) \left(\frac{C_p \eta_a}{k}\right)^{0.47} \left(\frac{D_t N'}{\bar{V}}\right)^{0.17} \left(\frac{D_t}{L}\right)^{0.37} \tag{4}$$

where h_p is the heat transfer coefficient of the sample film, which was scraped from the inside surface of the inner cylinder, N' is the rotational speed of the inner cylinder, k is the thermal conductivity of the sample, D_t is the inside diameter of the outer cylinder, and L is the length of each cylinder.

Figure 6.1 shows the correlation between the rheological constant K in Equation (1) and the heat transfer coefficient during the concentration of fruit juices. It follows from the results in Figure 6.1 that the shear thinning of fruit juices increases with progressive concentration.

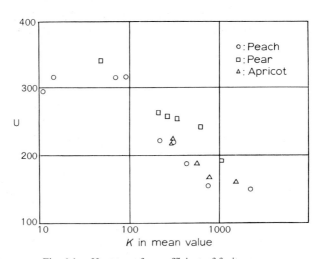

Fig. 6.1.   Heat transfer coefficient of fruit purees.

Figure 6.2 shows plots of Newtonian viscosity against heat transfer coefficient for water and honey. If the results in Figure 6.2 are compared with those in Figure 6.1, one finds higher values of the heat transfer coefficient for non-Newtonian fluids, so that shear thinning possibly plays a big part in the heat transfer phenomenon. For example, if one uses an ordinary type vacuum evaporator for concentration without any agitation, the heat transfer coefficient of the sample generally has a small value.

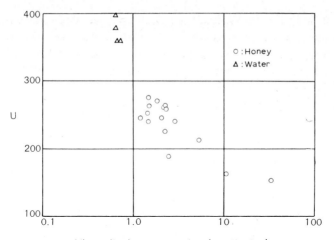

Viscosity in mean value (centipoise)

Fig. 6.2.   Heat transfer coefficient of honey and water.

6.2. Thermal Conductivity of Non-Newtonian Fluids

There have been few investigations of heat transfer phenomena during the steady-flow of non-Newtonian fluids. Charm and Merrill [4] derived an equation for the heat transfer coefficient of a pseudo-plastic body in a pipe, as follows;

$$\frac{hD}{k} = 2.0 \left(\frac{wC_p}{kL}\right)^{0.33} \left[\frac{K_a}{K_w}\left\{\frac{3 + 1/n}{(3 - 1/n)\,2}\right\}\right]^{0.14}, \tag{5}$$

where K_a and K_w are the pseudo-plasticity coefficient at the mean temperature of the system and at the wall temperature, respectively, h is the heat transfer coefficient of the system, D is the pipe diameter, k is the thermal conductivity of the system, w is the flux of the system, C_p is the specific heat of the system, L is the length of the pipe, and n is the power constant.

In the case of pseudo-plastic fluids, the following relation can be derived from the steady-flow equation in the pipe;

$$\frac{PD}{L\bar{V}^2 \varrho} = \frac{2^n 4K}{D^n \bar{V}^{2-n}}. \tag{6}$$

As it is possible to follow the relation $PD/L\bar{V}^2\varrho = 32/\mathrm{Re}$ for Newtonian fluids from dimensional analysis, the Reynolds number Re for non-Newtonian fluids can also be obtained from Equations (5) and (6), as follows;

$$\mathrm{Re} = \frac{8 D^n \bar{V}^{2-n}\varrho}{(1/n + 3)^n \, 2^n K}, \tag{7}$$

where the term $PD/L\bar{V}^2\varrho$ is quoted as a frictional coefficient.

Figure 6.3 shows plots of frictional coefficient against Reynolds number, though this correlation is not always useful for pseudo-plastic fluids or dilatant fluids. Metzner and Reed [5] have reported that the fluids showing abnormal flow generally become turbulent when the Reynolds number is greater than 70000.

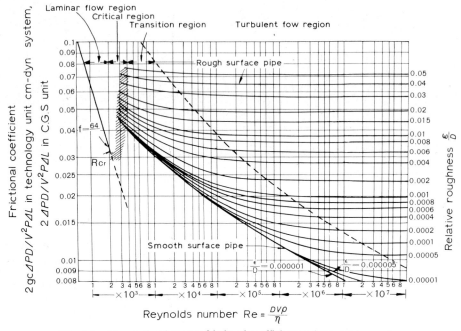

Fig. 6.3.   Correlation between frictional coefficient and Reynolds number.

Metzner has proposed the following equation for the pressure loss of pseudo-plastic fluids in cylindrical pipes due to turbulent flow

$$\frac{1}{\sqrt{\dfrac{DP/4L}{\varrho \bar{V}^2/2}}} = \frac{4.0}{n^{0.75}}\, \log\left\{\left(\frac{D^n \bar{V}^{2-n}\varrho}{K8^{n-1}}\right)\left(\frac{DP/4L}{\varrho\bar{V}^2/2}\right)^{1-n/2}\right\}. \tag{8}$$

6.3. Mixing of Non-Newtonian Fluids

Mixing and agitation are common techniques for processing foods. Preparation of emulsions or suspensions from the constituent ingredients requires the application of strong agitation to the system. The necessary motive force for agitation of non-Newtonian system is quite different to that for Newtonian fluids.

Mixing or agitation of liquids by means of a propeller is characterized by a mean rate of shear $(\dot{\gamma})_{ave}$. As the rate of shear is proportional to the rotational speed N of the propeller, the mean rate of shear can be expressed by

$$(\dot{\gamma})_{ave} = kN. \tag{9}$$

In order to obtain the proportionality constant k in Equation (9), it is necessary to measure the so-called 'power number' P_r/N^3D^4, where P_r is the motive force, and D is the diameter of the propeller. If the Reynolds number $\varrho ND^2/\eta_a$ can be obtained from Figure 6.3, one may calculate the apparent viscosity η_a of the system. Then, if one can derive the flow curve of the system, it should be possible to follow the mean rate of shear $(\dot{\gamma})_{\text{ave}}$, and, therefore, the proportionality constant k can be calculated from Equation (9).

6.4. Application to Coating

Coating techniques are sometimes required for preparing foods, e.g. chocolate coating on ice cream or on toffee, and the thickness of the coating on the food is affected by the consistency of the coating material.

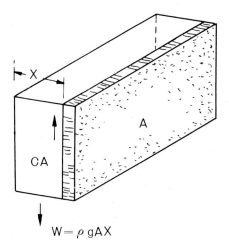

$$W = \rho\, g A X$$

Fig. 6.4.   Coating on to a vertical plane.

Figure 6.4 shows schematically the process of coating on to a vertical plane, in which the shearing stress appears in an interface between A (food base) and CA (coating material). Therefore, the following relation can be obtained

$$\varrho g_c X = C, \tag{10}$$

where ϱ is the density of coating material, g_c is the gravitational constant, X is the thickness of the coated layer, and C is the yield value of the coating material.

Equation (10) has been examined [7] for two kinds of chocolate with different fat contents, as shown in Table 6.1, and one can see that there is good agreement between the calculated and the experimental values. This result also suggests that a thin coating layer can be obtained using a fat-rich chocolate under suitable operating conditions.

When the coating material is spread onto a horizontal plane, the thickness of the coated layer h is affected by factors such as the normal force S_t, yield value of the coat-

TABLE 6.1
Thickness of chocolate on the coated layer [7]

Fat content in chocolate (%)	Yield value of chocolate C (dyn cm^{-2})	Thickness (cm)	
		Calculated	Observed
31	205 (36.6 °C)	0.146	0.144
47	42 (36.6 °C)	0.0310	0.0311

ing material C, cumulative weight W of the coating material, contact angle θ to the plane, etc., as follows:

$$h = \frac{4\pi S_t^2 \varrho^2}{W (\varrho g_c)^3 (\sin \theta)^2},$$ (11)

where the normal force S_t can also be defined by

$$S_t = \frac{W}{2\pi (w/h\varrho g_c \pi)^{1/2} (C/\varrho g_c \sin \theta)},$$ (12)

where w is the weight of coated layer per unit area.

6.5. Measurements of Pipe Flow Flux of Non-Newtonian Foods

Eolkin [8] has tried to design an apparatus for continuously measuring the consistency of foods. His apparatus is shown in Figure 6.5, where A and D are thin pipes which are connected to thick pipes B and C, respectively. The symbols P_1, P_2, P_3 and P_4 in Figure 6.6 represent the pressure at each point. If a Newtonian fluid flows in this apparatus, the flux per unit time V/t in the thin pipe can be expressed by the equation

$$V/t = \pi P r^4/8 \cdot l\eta,$$ (13)

where P is the initial pressure to the fluid, r is the radius of the pipe, l is the length of the pipe, and η is the Newtonian viscosity of the fluid. Let us suppose that the pressure loss of the fluid in pipes A and B is $\Delta P'$, and that the pressure loss in C and D is $\Delta P''$, then the following relation can be derived

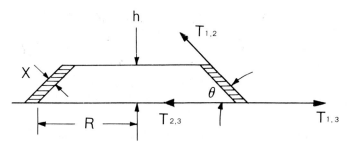

Fig. 6.5. Coating on to a horizontal plane.

$$\Delta P' = (P_2 - P_3) - (P_1 - P_2) \tag{14}$$

$$\Delta P'' = (P_1 - P_4) - (P_4 - P_3) \tag{15}$$

From these equations it follows that for Newtonian fluids

$$P_1 - P_2 = P_2 - P_3 = P_1 - P_4 = P_4 - P_3, \quad \text{and} \quad \Delta P' = \Delta P'' = 0.$$

Therefore, no pressure will be registered in the differential manometer of the apparatus. However, in the case of non-Newtonian fluids, the consistency is affected by the rate of shear, so that

$$P_1 - P_2 < P_2 - P_3, \quad \Delta P' > 0$$
$$P_1 - P_4 > P_4 - P_3, \quad \Delta P'' < 0$$

and, accordingly,

$$P_1 - P_2 = P_4 - P_3, \quad P_2 - P_3 = P_1 - P_4.$$

Thus, it is possible to derive the following relations,

$$\Delta P' = (P_2 - P_3) - (P_4 - P_3) = P_2 - P_4 \tag{16}$$

$$\Delta P'' = (P_2 - P_3) - (P_4 - P_3) = P_2 - P_4 \tag{17}$$

and then, $\Delta P' = \Delta P''$. It follows from the above relations that the differential manometer will register a pressure difference between P_2 and P_4 for non-Newtonian fluid systems.

The pressure difference $P_2 - P_4$ is proportional to the viscosity difference under a

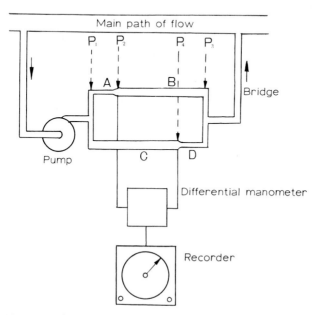

Fig. 6.6.   Schematic diagram of apparatus for measuring consistency.

fixed set of conditions; this means that the pressure difference corresponds to the shear rate gradient in the system. Figure 6.7 shows the characteristics of the apparatus, and the oblique lines indicate the region of high measurement sensitivity.

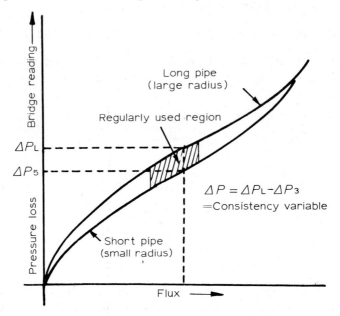

Fig. 6.7.   Characteristic curve of measurement with apparatus.

References

[1] Harper, J. C.: 1960, *Food Technol.* **14**, 557.
[2] Pigford, R. L.: 1955, *Chem. Eng. Prog. Symposium*, Ser. No. 17, 51 and 79.
[3] Skelland, A. H.: 1958, *Chem. Eng. Sci.* **7**, 166.
[4] Charm, S. E. and Merril, E. W.: 1959, *Food Res.* **24**, 319.
[5] Metzer, A. B. and Reed, J. C.: 1955, *Amer. Chem. Eng. J.* **1**, 434.
[6] Metzner, A. B.: 1958, *Rheolog. Acta* **1**, 205
[7] Charm, S. E.: 1963, *The Fundamentals of Food Engineering*, p. 101, Avi Pub. Co., Inc.
[8] Eolkin, D.: 1957, *Food Technol.* **11**, 254.

GENERAL BIBLIOGRAPHY

Although many books have been published on the rheology of synthetic macromolecules, we have very few books which cover systematically the subject matters on the consistency, texture and rheology of food-stuffs. Recently, however, it seems that the rheological studies of foodstuffs have come into the limelight, and it has been possible to list some unique and excellent books in this field. The author recommends some of them to the advanced scientist and technologist for further reading.

Charm, S. E.: 1963, *The Fundamentals of Food Engineering*, The AVI Publishing Co., Inc., Westport, Connecticut.

This book fully covers the subjects of food technology such as mass and heat transfers, thermal conductivity, fluid mechanics, etc. In view of rheology, viscosity is major subject, and it is much applicable to the food processings.

Charm, S. E.: 1962, 'The Nature and Role of Fluid Consistency in Food Engineering Applications', *Advan. Food Res.* **11**, 356.

Viscous non-Newtonian fluid foods are explained precisely in the light of the food processings. Most of the pages are well worth reading for food technologist.

Harper, R.: 1952, 'Psychological and Psycho-Physical Studies in Craftsmanship in Dairying', *British J. Psychol. Monograph Suppl.*, Cambridge University Press, London.

This is an elementary introduction of psycho-physics described from the view point of the relationship between the sensory evaluation of craftsmen in dairying and the rheological measurements. There are some difficult passages.

Kramer, A. and Twigg, B. A.: 1959, 'Principles and Instrumentation for the Physical Measurement of Food Quality with Special Reference to Fruit and Vegetable Products', *Advan. Food Res.* **9**, 153.

The principle and method for measuring the texture of fruits and vegetables are described together with rheological aspect of the shear pressure.

Nakagawa, T.: 1966, *Handbook of Food Chemistry* (in Japanese), Asakura Publishing Co., Ltd., Tokyo.

The author is famous for research activities in the field of rheology and polymer chemistry. He explains the principle and method of food rheology in this book using a classification of foods such as viscous foods, pasty foods, semisolid foods, etc.

Matz, S. A.: 1962, *Food Texture*, The AVI Publishing Co., Inc.

The source of this book consists of review papers written by thirty-one food scientists. First and second chapters; the basic back-ground of food texture and classification of the subject matters, third chapter; interrelationship between food processing and texture, fourth chapter; change of texture during storage. One may be suggested various problems on food rheology from this book.

Scott Blair, G. W.: 1969, *Elementary Rheology*, Academic Press, London and New York (translated in Japanese by Oka, S. and Azuma, T., Asakura Publishing Co., Ltd., Tokyo).

The author has much contributed to the development of rheology through his outstanding research works, and published a number of famous books in this field. He tries to explain an introduction of rheology in this book using a minimum of mathematics and intimate materials including foodstuffs.

Scott Blair, G. W. and Reiner, M.: 1957, *Agricultural Rheology*, Routledge & Kegan Paul, London (translated in Japanese by Sudo, S. and Yasutomi, R., Misuzu Publishing Co., Ltd., Tokyo).

Rheological phenomena in agriculture are well arranged in this book; people who work for agricultural science and technology should be suggested rheological idea to the agricultural subjects.

Scott Blair, G. W. (ed.): 1953, *Foodstuffs; Their Plasticity, Fluidity and Consistency*, Interscience Publishers, New York (translated in Japanese by Nikuni, J. and Isemura, T., Asakura Publishing Co., Ltd., Tokyo).

This is monographs on rheology. Materials dealt with are cereals, dairy products, jellies, syrups, confectionery, etc. Very unique part in this book is the description on psychorheology of foodstuffs.

Scott Blair, G. W.: 1958, 'Rheology in Food Research', *Advan. Food Res.* **8**, 1.

Mechanical properties of various foods, rheological phenomena in cooking, and psychorheology are well explained in this short paper from the view point of the simplified theories on rheology.

Szczesniak, A. S. and Torgeson, K. W.: 1955, 'Method of Meat Texture Measurement viewed from the Background of Factors Affecting Tenderness', *Advan. Food Res.* **14**, 33.

This paper covers the subject matters on the relationship between the texture of meat and the microstructural and biochemical changes of meat. The authors have studied the objective measurement of food texture at the Technical Centre, General Foods Co.

S. C. I. Monograph No. 7: 1960, *Food Texture*, Society of Chemical Industry, Belgrave Square, London.

Papers read at a Symposium on Texture in Foods in 1958, this consists of many interesting subject matters such as bread, chocolate, eggs, fish, etc.

S. C. I. Monograph No. 27: 1968, *Rheology and Texture of Foodstuffs*, Society of Chemical Industry, Belgrave Square, London.

Papers read at a Joint Symposium of the Society of Chemical Industry and the British Society of Rheology held in London in 1967, this has some interesting chapters on cake, bread, meat, fish, beans, potato, eggs, etc., together with papers on the psychophysical treatments of food texture.

Sherman, P. (ed): 1963, *Rheology of Emulsions*, Pergamon Press, Oxford.

The proceedings of a Conference of the British Society of Rheology held at Harrogate in 1962. One may find mention of emulsion studies through this book, and most of the chapters are well worth reading.

Sherman, P.: 1970, *Industrial Rheology*, Academic Press, London & New York.

The author is an expert of emulsion science, and he has done many studies on emulsion rheology. This unique book consists of six chapters, and two of them are on the rheological properties of foodstuffs, and on the correlation between the rheological properties and the sensory assessments of consistency.

Sherman, P.: 1968, *Emulsion Science*, Academic Press, London & New York (translated in Japanese by Sasaki, T., Hanai, T. and Mitsui, T., Asakura Publishing Co., Ltd., Tokyo).

Although one chapter in this book is only concerned with rheology, most of the contents are very useful, because emulsions resemble the dispersion state of foods.

Sone, T.: 1970, 'Rheology of Foods', in Oka, S. (ed.), *Elementary Rheology* (in Japanese), p. 51, Kogyo-Chosakai, Tokyo.

The author explains figuratively the principles of food rheology using frequent examples from occurrences in everyday life, e.g. ductility of caramel and metal, network structure and scaffold, etc.

Takano, T. and Sone, T.: 1961, *Flow and Transport in Food Industry* (in Japanese), Korin Publishing Co., Ltd., Tokyo.

This book covers theories and methods on rheology, and technological approaches to rheology in the field of food science, including descriptions on Kamaboko (fish jelly) and Udon (Japanese noodle).

The Society of Polymer Science, Japan: 1965, *Rheology Handbook* (in Japanese), Maruzen Publishing Co., Ltd., Tokyo.

This book is divided into two parts; one of them contains important data of rheological researches which have appeared in many scientific papers, and the other is a kind of dictionary for interpreting the rheological terminologies and meanings. This book is convenient for finding information on rheological subjects.

INDEX OF NAMES

INDEX OF SUBJECTS

Viscoelasticity 11, 23, 24, 75, 90, 92, 111, 163
Viscosity 7, 10, 11, 12, 35, 63, 84, 88
Viscous materials 11, 75, 77
Void per unit weight 41
Voigt-Kelvin's model 10, 11

Warner-Bratzler's testing machine 100, 101, 102
Weber-Fechner's law 159, 160

Weisenberg effect 61
Wheat flour 80, 112
– starch 71
Working 123, 152, 157

Yield value 7, 9, 29, 30, 114, 115, 123, 150, 152
Young's modulus 90